DIVINE PATTERNS

Unraveling the Mysteries of Scripture and Science

Robert E Johnson III

Divine Patterns
Copyright ©2024 by Robert E Johnson (DBA Three888s LLC)
Three888s Publishing
ISBN PB: 979-8-9916824-2-8

All rights reserved. No part of this book may be reproduced by any mechanical, photographic, or electronic process or in the form of photographic recording, nor may it be stored in a retrieval system, transmitted, or otherwise copied for public or private use without the prior written consent of the author/publisher, except for a reviewer who may quote brief passages for review.

The author and publisher assume no responsibility for errors or omissions or for damages resulting from the use of the information contained herein. The information contained in this publication is intended solely to provide general guidance on matters of interest for the personal use of the reader.

Special note: All conclusions derived from cited references in this book, references to other books, journal articles, periodicals, publications, website references, YouTube videos, encyclopedia references, or any other reference materials are those of the author and not intended to represent the views or findings of the original author or authors.

Book Cover Image by Freepik
Book design by Variance Author Services
www.varianceauthorservices.com

DIVINE PATTERNS

Unraveling the Mysteries of Scripture and Science

Robert E Johnson III

Acknowledgements

To my father, a WWII and Korean War veteran, who was providentially spared when members of his West Virginia Air National Guard Unit were killed in a 1948 plane crash. He was scheduled to be on that flight, but by God's grace, he missed it. Otherwise, I would not be here. I'm most grateful for his teachings, guidance, moral standards, encouragement, and constant push to excel at whatever I set my mind to.

To Dr. Ivan Panin for his amazing discovery of the numeric seal within the texts of the Old and New Testaments. I believe his discovery ranks alongside the unearthing of the Dead Sea Scrolls as one of the two most significant findings of the past 2,000 years regarding the reliability of the Scriptures.

To all the scientists, researchers, and apologists who have backed the reliability of God's word with compelling and overwhelming historical and scientific evidence.

To Variance Author Services and Tim Schulte for his helpful guidance in preparing and formatting the book for publication and for his assistance in designing the cover.

To Frank Roszell, Debbie Wooff, and Wade Harrison for their review of the book and recommended edits.

And last but certainly not least, to my wonderful family for their love, support, and encouragement, without which this book would not have been possible.

Contents

Introduction		1
1.	The Holy Bible – God's Message to Mankind	3
2.	The Divine Code in Scripture	6
3.	The Divine Code in the Book of Genesis	11
4.	The Divine Code in the Genealogy of Christ	17
5.	The Divine Code in the Book of Mark	21
6.	The Word of God – Logos 373	26
7.	Pi (π) and the Natural Algorithm (e)	31
8.	Salvation and the Number 8	37
9.	Sin and the Number 13	44
10.	Wisdom – The Foundation of God's Creation	56
11.	God's Seal in the Human Genome	60
12.	The Most Mysterious Number – 1/137	69
13.	Why are Fibonacci Numbers Everywhere?	75
14.	The Unspeakable Name	81
15.	Israel's Messiah	84
16.	The Names of Adam's Descendants	89
17.	The Shroud of Turin	92
18.	God Declares the End from the Beginning	99
19.	Conception and the Spark of Life	110
20.	Bible Truths That Predate Science	113
21.	God's Prophets Foretell the Future	121
22.	Dreams and Visions	139
23.	Logic and Reason	150
24.	The Laws of Thermodynamics	154
25.	Our Fine-Tuned Universe	157
26.	The Random Chance Hypothesis	164

27.	The Anthropic Principle	169
28.	Master Designer or Evolution	177
29.	Information Theory and DNA	185
30.	Evolution – Absence of Rational Answers	188
31.	Intelligent Design – Seed Dispersal Systems	192
32.	The Multiverse – A Search for Answers	197
33.	The Origin of Man – When Did We Get Here?	200
34.	The Earth's Age – Surprising Evidence	206
35.	Man's Science Versus God's Science	215
36.	What the Bible Says About Sin	224
37.	Fear of the Lord – The Foundation of Wisdom	232
38.	Two Worldviews and Their Implications	240
39.	Your Worldview Determines Your Destiny	243
40.	Rebellion – The World Accepts the Antichrist	249
41.	End of The Age – Judgment and Tribulation	252
42.	Quotes from the World's Brightest Minds	262

Works Cited   274

Author's Biography   308

Introduction

The purpose of this book is to present scientific, mathematical, and biblical evidence to prove the existence of God. This book will demonstrate that the original Hebrew and Greek texts of the Old and New Testaments are written with a divine seal that prevents alterations to the text without breaking the seal. The seal and the numeric patterns appear to be far too complex to have been conceived by the mind of man. This numeric seal is present in every book of the Bible and can even be found embedded in the names of Jesus, Satan, and the genetic code.

This book will also examine how God has used Fibonacci numbers throughout creation. What makes these numbers extraordinary is their highly unusual progression and their ability to produce optimization and efficiency in growth patterns.

We will also explore what physicists call the most mysterious number in the universe. It is a dimensionless number that appears at the intersection of relativity, electromagnetism, and quantum mechanics, and it is said to be the most fundamental unsolved problem in physics. This number, known as the fine structure constant, appears everywhere and is integral to the very structure of the universe. Astrophysicist Paul Sutter stated, "If it had any other value, life as we know it would be impossible. And yet we have no idea where it comes from." [1]

We'll also examine the anthropic principle, which postulates that the universe appears designed to support life. It explores in detail the mathematical probability that the cosmological constants observed in our universe could have occurred by chance. We will explore the scientific evidence for the age of the Earth, how long humans have existed, and why evolutionary theory is inadequate to explain how life could have evolved on a primordial Earth.

We will review the scientific analysis of "The Shroud of Turin" and present the first credible theory on how the image of Christ was transferred to this linen cloth, believed by many to be the actual burial cloth of Jesus Christ. There is a faint image of a crucified man embedded in the fabric. How the image got onto the cloth is one of the most perplexing and long-standing mysteries in modern science.

This book challenges many widely held beliefs. Among them are the reliability of the Bible, the trustworthiness of science regarding the origin and age of the universe, the plausibility of life on Earth emerging through a process called abiogenesis, and many other theories that directly contradict the word of God. The claims in this book are based on scientific evidence and the absolute reliability of God's word, which is based on historical facts, fulfilled prophecy, and a divine numeric seal embedded in the original Hebrew and Greek texts of the Bible.

- 1 -

The Holy Bible
God's Message to Mankind

Throughout human history, one book stands head and shoulders above all others. It has captured the hearts and minds of people from every race and nation. This book is the number one bestseller of all time. In 2021, research by the British and Foreign Bible Society found that the Bible has sold between five and seven billion copies. [2] What makes it so popular is its claim of divine origin, its profound wisdom, and its prophetic accuracy. The Bible presents a comprehensive history and detailed genealogy of mankind from the dawn of time. It foretells future events, provides guidance for living a purposeful life, and is filled with teachings and instructions for building an eternal relationship with the Creator of the universe.

But how can we be sure that every word in the Bible is true? How can we know that it isn't just the ideas and thoughts of the 40 writers who compiled it? How can we know that the Bible is God's revelation and not the opinion of men? Did the writers record God's general idea, or did they record His message word for word, letter for letter? Is it possible that they misinterpreted some of the message or changed the meaning as new translations were published? How can we know whether this book is the inspired word of God, as claimed in 2 Timothy 3:16?

Let's start by considering a few historical facts. This book was written before the invention of electricity, gasoline engines, airplanes, televisions, cell phones, space travel, the internet, and artificial intelligence. Yet, it accurately describes our current world. The Bible has revealed hundreds of events that have occurred on the world stage, years and sometimes centuries, in advance of their fulfillment. It accurately predicted the destruction of cities and nations, the birth of kings, and the rise and fall of world empires. It provides specific details of a future event that will be witnessed simultaneously by the entire world population. It describes a future financial system where individuals will require a mark on their hand or forehead to buy or sell. Remarkably, these predictions were written nearly two thousand years ago, long before the technology to make them possible existed. Today, the technological advancements to fulfill both prophecies are a reality.

How could the authors of the Bible have foreseen and written about events that would occur hundreds or even thousands of years in the future? Two thousand years before the invention of televisions, cell phones, or the internet, how could the writers of the Bible have known that the entire world population would simultaneously be able to witness a specific world event?

If you're searching for evidence that proves beyond a reasonable doubt the existence of God and the divine authorship of the Bible, this book provides the information you're looking for. It will reveal the existence of a divine seal, a numeric signature embedded in the original Hebrew and Greek texts of both the Old and New Testaments. The information that follows provides scientific evidence that the Bible is reliable, dependable, and accurate regarding creation, science, and past and future history.

My purpose in writing this book is to address any doubt regarding whether God, the Holy Spirit, is the actual author of the Bible. I will show that the writers of the 66 books inscribed His message exactly as given in the original Hebrew and Greek texts of the Old and New Testaments. I will explain how the Bible is unlike any secular or other religious book and how it has been encoded with a seal that ensures the accuracy of the original text. Understanding the significance and complexity of the encrypted seal that God used to authenticate the divine nature of the Bible is essential for comprehending everything else that follows in this book.

What I believe you will discover is that the seals and prophecies in the Old and New Testaments did not come from the minds of the writers. They were divinely inspired. [3] My hope is that the evidence you are about to encounter will prove beyond a reasonable doubt that the Bible could not have been written by mere men.

- 2 -

The Divine Code in Scripture

In the same way that governments use watermarks, security threads, and color-changing ink to prevent the counterfeiting of their currency, God has embedded a hidden security code in the original Hebrew and Greek texts of the Old and New Testaments of the Bible. The purpose of the code is twofold. The first is to verify the identity of the author, which is accomplished by the complexity of the code. The second is to preserve the accuracy of the text. Any changes to the original Hebrew and Greek texts, even individual letters, would destroy the code, the seal, and the divine pattern.

Amazingly, this security code, or "Seal of God," as some have called it, went undiscovered for over 4,000 years. The underlying seal is primarily heptadic, based on patterns of seven. However, other numeric features have also been discovered within the basic pattern, greatly increasing the overall complexity of the embedded code.

The individual who discovered the Bible's heptadic code was Dr. Ivan Panin. He was born in Russia on December 12, 1855. He immigrated to Germany in 1874 and then to the United States in 1877. He attended Harvard University, majoring in Greek and Hebrew, and earned a master's degree in Literary Criticism in 1882. [4] Ivan Panin was a college professor, a mathematician, and a linguist who traveled the country giving lectures on literary criticism. He was a well-known agnostic before making his discovery in 1890. [5]

Ivan Panin was likely introduced to the unique characteristics of the Hebrew and Greek alphabets during his studies at Harvard. The numeric values assigned to the letters of the Hebrew and Greek alphabets made them ideal platforms for the underlying security code that he discovered in the Holy Scriptures. What he ultimately found was an extremely complex and intricate numeric pattern of sevens, which ran from the very first sentence in the Book of Genesis to the last sentence in the Book of Revelation. He believed that the complexity of the numerical patterns he discovered was beyond the creative ability of human beings. He said that the more time he spent on a particular section of Scripture, the more patterns within patterns he found.

After making his discovery, Ivan Panin accepted Christ and dedicated the next 50 years of his life to putting the numeric values under the Greek and Hebrew letters in the Old and New Testaments and calculating the probability that they could have occurred by chance. He concluded that the likelihood of these patterns occurring by chance was virtually zero. He was convinced that the 40 authors of the Bible did not conspire over a period of 1,500 years to produce this heptadic code. He believed the Holy Spirit was the true author and that He dictated His message to human authors, who inscribed it perfectly, word for word, letter for letter.

A brief description of what Ivan Panin found was that the number of words in a particular passage of Scripture was divisible by seven; the letters were divisible by seven; the nouns were divisible by seven; the vowels and the consonants were divisible by seven; the number of words that started or ended with a consonant or vowel was divisible by seven. Often, the features of seven went well beyond these examples. He frequently calculated the likelihood that the patterns he discovered could have occurred by chance and found the probabilities to be more than trillions or quadrillions to one.

Not one letter (jot or tittle) is missing from the original text; otherwise, the pattern of sevens he discovered would disintegrate. For example, he found that the removal of a single letter in the original text could change the number of letters, the number of vowels, the number of consonants, the words that started or ended with a consonant, the words that started or ended with a vowel, the words that appeared once or more than once, and the value of the passage would no longer be divisible by seven.

Over his lifetime, Ivan Panin compiled over 40,000 pages of his research and calculations and presented them to the Noble Research Foundation, claiming that his discovery provided compelling evidence that the Holy Scriptures were divinely inspired. I believe Ivan Panin's discovery of the numeric seal within the text of the Bible, along with fulfilled prophecies, provides the most compelling evidence that God is the author of this divine book.

Ivan Panin knew that, unlike the English alphabet, the letters in the Greek and Hebrew alphabets have numeric values, and both alphabets use a similar method for assigning those values. The first ten letters in both alphabets have the values 1 through 10, and the next nine letters have values that are multiples of ten, i.e., 20, 30, 40, etc., until reaching the 19th letter, which has a value of 100. The remaining letters are in multiples of a hundred, i.e., 100, 200, 300, etc. The following tables show the numeric values of the letters of the Hebrew and Greek alphabets.

THE 22 HEBREW LETTERS AND THEIR NUMERICAL VALUES											
1	2	3	4	5	6	7	8	9	10	11
א	ב	ג	ד	ה	ו	ז	ח	ט	י	כ
1	2	3	4	5	6	7	8	9	10	20
Alef	Bet	Gimel	Dalet	Hey	Vav	Zayin	Chet	Tet	Yod	Kaph
12	13	14	15	16	17	18	19	20	21	22
ל	מ	נ	ס	ע	פ	צ	ק	ר	ש	ת
30	40	50	60	70	80	90	100	200	300	400
Lamed	Mem	Nun	Samech	Ayin	Pe	Tsade	Qoph	Resh	Shin	Tav

THE 27 GREEK LETTERS AND THEIR NUMERICAL VALUES							
α	Alpha	1	ι	Iota	10	ρ	Rho	100
β	Beta	2	κ	Kappa	20	σ	Sigma	200
γ	Gamma	3	λ	Lambda	30	τ	Tau	300
δ	Delta	4	μ	Mu	40	υ	Upsilon	400
ε	Epsilon	5	ν	Nu	50	φ	Phi	500
ς	Stigma	6	ξ	Xi	60	χ	Chi	600
ζ	Zeta	7	ο	Omicron	70	ψ	Psi	700
η	Eta	8	π	Pi	80	ω	Omega	800
θ	Theta	9	ϙ	Koppa	90	ϡ	Sampi	900

The following chapters will demonstrate how the Old and New Testaments have been encoded primarily with a seal of sevens. Unlike the English alphabet, the Hebrew and Greek alphabets have numeric values assigned to their letters. This feature makes them uniquely suitable for embedding the heptadic seal that distinguishes the Bible from all other written works. It's interesting to note that the two alphabets that God used to communicate His message to mankind have a total of 49 letters between them. The Hebrew alphabet has 22 letters, and the Greek alphabet has 27. The first occurrence of the divine seal of sevens appears in the number of letters in the alphabets used to write the texts, as the factors of 49 are (7 x 7).

- 3 -

The Divine Code in the Book of Genesis

Arguably, Genesis 1:1 is the most amazing sentence ever written. The level of scientific information, numerical coding, and divine revelation contained in this single sentence is simply mind-boggling. Dr. Ivan Panin's discovery of a pattern of sevens within these divinely inspired words provides but a glimpse of the unique characteristics and hidden messages within this remarkable sentence.

In the original Hebrew, there are seven words in Genesis 1:1. They are translated in English as "In the beginning, God created the heavens and the earth." The sum of the numeric values of the letters in a word, or the sum of the numeric values of the words in a sentence, is referred to as the standard gematria value. The gematria value of each Hebrew or Greek word, sentence, or passage is calculated by summing the values of their individual letters. Below are the seven Hebrew words in Genesis 1:1 and their corresponding numeric values. The first word, which translates as "In the beginning," has a value of 913. Note that Hebrew is read from right to left.

הארצ	ואת	השמימ	את	אלהימ	ברא	בראשית
the earth	and	the heavens	(untranslated)	God	created	In the beginning
296	407	395	401	86	203	913

In the original Hebrew, Genesis 1:1 has the following features of the number seven.

There are 7 Hebrew words	7 x 1
There are 28 total letters	7 x 4
The three nouns have a value of 777	7 x 111
The three nouns have 14 letters	7 x 2
The 3rd and 4th words have 7 letters	7 x 1
The 4th and 5th words have 7 letters	7 x 1
The 6th and 7th words have 7 letters	7 x 1
The first three words have 14 letters	7 x 2
The last four words have 14 letters	7 x 2
The value of the verb "created" = 203	7 x 29
The last letters in the 1st and last word = 490	7 x 70
1st and last letters of 1st and 7th words = 497	7 x 71
1st and last letters of words 2, 3, 4, 5, & 6 = 896	7 x 128
1st and last letters of words 1, 2, & 3 = 42	7 x 6
1st and last letters of words 4, 5, 6, & 7 = 91	7 x 13
1st and last letters of all 7 words = 1,393	7 x 199

[6]

These 16 features of seven in Genesis 1:1 are extraordinary examples of God's heptadic seal on the Holy Scriptures, and it is by no means a complete list. But, in addition to the above gematria values of 7, there are more than 21 additional features of the number 37. We will examine the features of 37 in Genesis 1:1 later in a chapter titled "The Logos 373 – The Word of God."

Seven is frequently referred to as the number representing perfection or completeness. It is the number most associated with God the Father. No other number appears more frequently in Scripture. Ivan Panin believed that God used patterns of seven to authenticate or counterfeit-proof the 66 books of the Old and New Testaments. The number seven appears 735 times in Scripture, and 735 is itself a multiple of 7 (7 x 7 x 15). [7]

Here are some additional examples of God's use of this profoundly significant number in Scripture. There were seven days in the creation event, the seventh being a day of rest. God ordained that the Hebrew calendar would be based on a seven-day week. God gave Israel seven feasts to observe and celebrate (Passover, Unleavened Bread, First Fruits, Pentecost, Trumpets, Atonement, and Tabernacles). God told Noah to take seven pairs of each clean animal with him on the ark. In Matthew 18:2, Peter asked Jesus, "Lord, how many times shall I forgive my brother or sister who sins against me? Up to seven times?" Jesus answered, "I tell you, not seven times, but seventy times seven." In the Book of Revelation, Jesus speaks of seven candlesticks, seven spirits, seven churches, seven angels, seven stars, seven seals, seven trumpets, seven thunders, seven last plagues, and seven bowls. The period of tribulation is prophesied to last seven years. Since seven is the most frequently appearing number in Scripture, is it surprising that His Word would also be sealed with this divine number?

As seven is the number representing perfection or completion, is it coincidental that Jesus, the perfect sacrificial Lamb of God, would complete His mission on Earth with seven final pronouncements from the cross? Below are the final words that Jesus spoke before dying on the cross.

1. To his persecutors, He said, "Father, forgive them, for they do not know what they do." (Luke 23:34 NIV)

2. To the thief on the cross, He said, "I tell you the truth, today you will be with me in paradise." (Luke 23:43 NIV)

3. To his mother and John, He said, "Woman, behold your son!" Then He said to the disciple, "Behold your mother." (John 19: 26-27 NKJV)

4. To God the Father, He said, "My God, my God, why have You forsaken me?" (Matt 27:46 NIV)

5. To fulfill the prophecy in Psalm 69:21, He said, "I thirst!" (John 19:28 KJV)

6. To signify His mission on Earth was complete, He said, "It is finished!" (John 19:30 KJV)

7. To indicate His restoration to the Father, He said, "Father, into your hands I commit my spirit." (Luke 23:46 NIV)

Another feature of seven that may have been overlooked relates to the number of books in the Old and New Testaments. The prevailing view of most Christians is that the Bible is composed of 66 books, but it is actually composed of 70 books (7 x 10). For some reason, a decision was made to count the five books of Psalms as a single book. However, Psalms is composed of five books. Psalms 1 through 41 is Book I; Psalms 42 through 72 is Book II; Psalms 73 through 89 is Book III; Psalms 90 through 106 is Book IV; and Psalms 107 through 150 is Book V. We don't count I Samuel and II Samuel as one book, or I Kings and II Kings as one book, or I John, II John, and III John as one book, so why do we count the five books of Psalms as one book? If all five books in Psalms were counted as separate books, the total number of books would be 70. Not surprisingly, this number is consistent with the seal of seven (7 x 10) seen throughout this divinely inspired book.

Consider the fact that the books of the Bible were written over a period of about 1,500 years by 40 different writers and that the numeric patterns found in the Bible were only discovered in the last 130 years. Since all 66 or 70 books, depending on how you count, exhibit these same numeric features of seven, the 40 authors would have had to conspire with one another to ensure they all used the same heptadic formula, i.e., a pattern of sevens. This would have been impossible given that their lives spanned a period of more than 20 generations. They would also have had to know the values of the letters in the Hebrew and Greek alphabets. However, that would have been impossible since the values for the letters of the Hebrew alphabet had not yet been determined at the time of the writings of the Old Testament. "The first evidence of the use of Hebrew letters as numbers dates to 78 BCE." [8] Besides, even if they had known the values of the letters in their alphabets, the complexity of writing coherent sentences and keeping the overall theme of the Bible intact would have been beyond their capability.

– 4 –

The Divine Code in the Genealogy of Christ

Another example of God's divine code or fingerprint in the Holy Scriptures is found in the genealogy of Christ, which is recorded in the first chapter of the Book of Matthew. Here again, we find the seal of seven embedded in the text. When considering whether the writer manipulated the names to achieve this pattern of sevens, keep in mind that the names in the genealogy were all recorded before Matthew was born. Therefore, it would have been impossible for Matthew to have chosen names that would produce these features of seven. Is it possible that God prompted the parents of these individuals to use those names so that a pattern of sevens would be produced as evidence of His divine providence in the affairs of men?

Below are the features of seven that Ivan Panin discovered in Matthew 1, the first book of the Greek New Testament.

The genealogy has three sets of 14 generations	7 x 6
Abraham to David – 14 generations	7 x 2
David to the exile in Babylon – 14 generations	7 x 2
Exile in Babylon to Christ – 14 generations	7 x 2
The number of words in the passage is 49	7 x 7

Of these 49 words, 28 begin with a vowel	7 x 4
Of these 49 words, 21 begin with a consonant	7 x 3
Of these 49 words, 7 end with a vowel	7 x 1
Of these 49 words, 42 end with a consonant	7 x 6
Of these 49 words, 35 occur more than once	7 x 5
14 of the 49 words occur only once	7 x 2
42 of the 49 words occur in one form	7 x 6
7 of the 49 words occur in more than one form	7 x 1
42 of the 49 words are nouns	7 x 6
7 of the 49 words are not nouns	7 x 1
These 49 words contain 266 Greek letters	7 x 38
Of these 266 Greek letters, 140 are vowels	7 x 20
Of the 266 Greek letters, 126 are consonants	7 x 18
Of the 42 nouns, 35 are proper nouns	7 x 5
The 35 proper nouns appear 63 times	7 x 9
The number of male names is 28	7 x 4
These male names appear 56 times	7 x 8
The three women's names have 14 letters	7 x 2
The number of compound nouns is 7	7 x 1
The seven compound nouns have 49 letters	7 x 7

[9]

Note: The probability of getting 25 features of the number seven by random chance is one chance in 1,341,068,619,663,964,900,807, or one chance in 1.3 Sextillion!

These numeric features, or patterns of seven, provide undeniable evidence that the Bible has been written with an extremely sophisticated heptadic code. Ivan Panin claims to have found this code embedded in every book of the Bible. He believed the complexity of the code was beyond the ability of humans to construct. If he is correct, and I'm convinced that he is, his discovery provides compelling evidence that the Holy Spirit is the true author of this divine book. Ivan Panin believed that no human author could have produced the astronomical improbabilities that he found throughout the entire text of the Bible.

The late Dr. Chuck Missler was very familiar with Ivan Panin's discovery. Dr. Missler founded Koinonia House Ministries and was a renowned Bible scholar. He received an honorary Doctor of Divinity degree from Louisiana Baptist University. He earned a Bachelor of Science degree in Engineering from the United States Naval Academy and a master's degree in engineering from UCLA. He also pursued postgraduate studies in advanced mathematics at UCLA. In the corporate world, he served as chairman and CEO of Western Digital, chairman of Resdel Industries, and chairman of Phoenix Group International. He also served on the Board of Directors for the Computer and Communications Industry Association in Washington, D.C. Due to Dr. Missler's stellar reputation, he was invited to participate in a highly sensitive advanced technology project with William E. Simon, former Secretary of the Treasury; General David C. Jones, Chairman of the Joint Chiefs of Staff; and Dr. Edward Teller, the scientific advisor to the President of the United States. [10]

Dr. Missler's education and background are provided above to lend credence and add weight to his statement regarding the difficulty and complexity of producing the 75 features of seven that Ivan Panin found in the last 12 verses of the Book of Mark. Here is what Dr. Missler had to say regarding those verses: "If a supercomputer could be programmed to attempt 400 million tries per second, it would take 1,000,000 of them over 4,000,000 years to produce a combination of these 75 heptadic features by chance." [11] If this is how long it would take a million supercomputers to produce 75 features of seven by chance, what's the likelihood Mark produced them on his own within a few years or months?

Dr. Missler's remarks may help explain why Ivan Panin believed that the complexity of the numeric patterns he found from the Book of Genesis to the Book of Revelation could not have been constructed without the guidance of the Holy Spirit. Ivan Panin was so convinced that this code could not be replicated by humans that he offered a large sum of money to anyone who could write 300 words that conveyed a coherent message that duplicated the numeric patterns he found in the Bible. He stated that those competing for the prize could assign any values to the letters in our alphabet as long as the values remained consistent. Not surprisingly, no one was able to duplicate the level of complexity that Ivan Panin found in the Holy Scriptures.

– 5 –

The Divine Code in the Book of Mark

The first eight verses in the Book of Mark describe the ministry of John the Baptist. Once again, embedded in these verses is the seal of seven, God's fingerprint, which ensures the accuracy of the Scriptures. The following features of seven are found in the original Greek text in Mark 1:1-8 and were discovered by Ivan Panin.

The total words in the passage are 126	7 x 18
The total vocabulary words are 77	7 x 11
The letters in the vocabulary words are 427	7 x 61
The vowels in the 427 letters are 224	7 x 32
The consonants in the 427 letters are 203	7 x 29
The vocabulary words spoken by John are 21	7 x 3
The remaining vocabulary words are 56	7 x 8
Of the 77 words, 42 begin with a vowel	7 x 6
Of the 77 words, 35 begin with a consonant	7 x 5

Note: The probability that nine features of seven would occur by random chance is one in 40,353,607. However, this number is dwarfed by the probability of the number of features of seven in the last 12 verses of the Book of Mark.

In most Bibles, there is a footnote following the last 12 verses in the Book of Mark. The footnote states that these verses were not included or found in the two earliest known manuscripts of the New Testament. However, they were found in hundreds of 1st-century Bibles. The question is: Should they have been included? Did the Holy Spirit inspire them, or were they simply words written by Mark? If the Holy Spirit inspired them, they should have His seal, which would authenticate them. If the seal is not there, they should not have appeared in the New Testament since they were not inspired by God.

The last 12 verses in the Book of Mark include the following words spoken by Jesus to his disciples after his resurrection: "'And He said unto them, 'Go ye into all the world and preach the gospel to every creature. He who believes and is baptized will be saved; but he who does not believe will be condemned. And these signs will follow those who believe: In My name they will cast out demons; they will speak with new tongues; they will take up serpents; and if they drink anything deadly, it will by no means hurt them; they will lay hands on the sick, and they will recover.'" (Mark 16:15-18 NKJV)

Since God inspires all Scripture, if these words were inspired, they would have God's divine seal. What Dr. Ivan Panin discovered in these verses was mathematically astounding. He found 75 features of the number seven. He even published a book on his findings, which clearly shows that these verses had to have been inspired. The probability that these verses occurred by chance is 7 to the 75th power or one chance in 2,411,865,032,257,060,000,000,000,000,000,000,000,000,000,000,000,000,000.

To understand the enormity of the above number, let me attempt to put it into perspective. There are 500 billion trillion atoms in a teaspoon of water. [12] The above number is more than a trillion times the number of atoms in the entire Earth—meaning the likelihood that the last 12 verses in the Book of Mark occurred by chance would be the same likelihood as marking one atom and blindly selecting it from all the atoms in a trillion different Earths. These 75 features of seven did not occur by coincidence, and Mark would not have been able to construct these features on his own, even if he had access to supercomputers with artificial general intelligence (AGI).

Listed below are just 25 of the 75 examples of seven that Ivan Panin found in Mark 16:9-20. They were the last words Jesus spoke to His disciples before ascending into Heaven. They are referred to as "The Great Commission."

The 12 verses of Mark contain 175 words	7 x 25
Value of the entire passage = 103,663	7 x 14,809
The value of all word forms = 89,663	7 x 12,809
The total unique vocabulary words are 98	7 x 14
The unique vocabulary has 553 letters	7 x 79
The unique vocabulary has 294 vowels	7 x 42
The number of consonants is 259	7 x 37
84 of 98 are found elsewhere in Mark	7 x 12
14 of 98 are only found here	7 x 2
Lord's speech, verses 15-18, has 42 words	7 x 6
The remaining words are 56	7 x 8
The number of word forms is 133	7 x 19

112 of the word forms occur only once	7 x 16
21 of word forms occur more than once	7 x 3
The 21 word forms have 231 letters	7 x 33
The value of all the word forms = 86,450	7 x 12,350

The natural divisions of the whole passage are (v. 9-11), Jesus' appearance to Mary Magdalene, (v. 12-14), Christ's appearance to two of them as they walked, (v. 15-18), the speech of Christ, and (v. 19-20), the conclusion. The gematria or numeric values of the letters for these verses are shown below.

Verses 9-11 have 35 words	7 x 5
Verses 12-18 have 105 words	7 x 15
Verses 19-20 have 35 words	7 x 5
Letter values in verses 9-11 = 17,213	7 x 2,459
Letter values in verses 9 and 11 = 11,795	7 x 1,685
Letter values in verse 10 = 5,418	7 x 774
Value of the first word in verse 10 = 98	7 x 14
Middle word value in verse 10 = 4,529	7 x 647
Last word value in verse 10 = 791	7 x 113

In addition to the above findings, Ivan Panin also found similar examples of seven in the numeric values of the vocabulary words and their word forms. It would have been virtually impossible for human authors to produce these highly complex numeric patterns thousands of years before the advent of computers.

It's worth repeating Dr. Chuck Missler's comments regarding these 75 features of seven in Mark 16:9-20. He said, "If a supercomputer could be programmed to attempt 400 million tries per second, it would take 1,000,000 of them over 4,000,000 years to produce a combination of 75 heptadic features by chance." He went on to say, "Just as we encounter coding devices in our high technology environments, here we have an automatic security system that monitors every letter of every word, that never rusts or wears out and has remained in service for almost two thousand years! It is a signature that can't be erased and that counterfeiters can't simulate. Why are we surprised? God has declared that He magnified His Word even above His name!" [13]

– 6 –

The Word of God
Logos 373

The number 37 has been interpreted to mean the "Word of God." John 1:1 says, "In the beginning was the Word, and the Word was with God, and the Word was God." The "Word" in English is a translation of the Greek word Logos. The Greek spelling is λογοδ. These five Greek letters are Gamma, Omicron, Lambda, Omicron, and Sigma. The gematria values for these letters total 373. Gamma = 3, Omicron = 70, Lambda = 30, Omicron = 70, and Sigma = 200. Note that the number 37 in the gematria of Logos (373) appears in the first two digits, whether looking forward or backward. The gematria, or numeric value, of all seven words in Genesis 1:1 is 2,701 or (37 x 73). Once again, we see the number 37 in the factors of 2,701, whether looking forward or backward. This number 37 identifies, or proclaims numerically, the identity of the Creator: He is the Logos, the Word of God, the one who created the heavens and the earth, and His number appears multiple times in Genesis 1:1. We read in Colossians 1:15 (NKJV), "He (Jesus) is the image of the invisible God, the firstborn over all creation." Is it possible that this idea, Jesus being the image of God, is also being expressed in the underlying numerics of Genesis 1:1. Is it possible that the number 37 is also a reflection of God the

Father? Is it possible His number is 73? Seven is the number most frequently associated with God. Could 73 represent God (7) in (3) persons?

In addition to 37 being the number for the Word, it is also embedded in many of Jesus' other names. Some examples include Jesus – gematria 888 (37 x 24); Christ – gematria 1,480 (37 x 40); and Son of Man – gematria 2,960 (37 x 80).

Both 37 and 73 are prime numbers, and their product 2,701, the gematria value of Genesis 1:1, is the 37th hexagonal number and the 73rd triangular number. [14] [15] Are all these numerics just coincidences, or is God the Son being revealed as the image of the Father?

The Bible suggests that God has a special and unique relationship and love for the Jewish people. We read in Deuteronomy 7:6-8 (NKJV), "For you are a holy people to the Lord your God; the Lord your God has chosen you to be a people for Himself, a special treasure above all the peoples on the face of the earth. The Lord did not set His love on you nor choose you because you were more in number than any other people, for you were the least of all peoples; but because the Lord loves you, and because He would keep the oath which He swore to your fathers..." Is it not interesting that the factors of the gematria of Genesis 1:1 (37 and 73) are the only two numbers between 1 and 100, besides 13, that can form a perfect six-pointed Star of David by arranging 37 or 73 dots or stars? See the 13 stars arranged in the form of a Star of David above the eagle on the back of a one-dollar bill.

This number 37 also shows up in the second half of the second sentence of Genesis 1:2 in the words, "...And the Spirit of God moved upon the face of the waters." The numeric value of the letters in this phrase is 1,339 (37 x 37). Is it likely that all these numerical improbabilities are due to random chance?

If Jesus is the Word, the Logos, the one who made all things that have been made, is it surprising to find his number 37 in the sentence that describes the creation of heaven and earth? Is it a coincidence that the product of 37 and 73 (the reflection of 37) is the sum of the seven words in Genesis 1:1? Is it just a coincidence that the number 37 is seen forward and backward in the Greek word for Logos and the factors of the gematria for Genesis 1:1 (37 x 73)? Is it possible that these patterns were embedded in the Holy Scriptures as numeric evidence or clues regarding the identity of the Creator? Let's examine the features of just the number 37 in Genesis 1:1: "In the beginning God created the heavens and the earth." The numeric value or gematria of the original Hebrew words follow.

בראשית	ברא	אלהימ	את	השמימ	ואת	הארצ
In the beginning	created	God	(untranslated)	the heavens	and	the earth
913	203	86	401	395	407	296

The following 21 features of the number 37 are embedded in the above seven words from Genesis 1:1.

The value of words 1 and 3 is 999	37 x 27
The value of words 1, 3, and 6 is 1406	37 x 38
The value of words 1, 3, and 7 is 1295	37 x 35
The value of words 1, 3, 6, and 7 is 1702	37 x 46
The value of words 1, 2, and 4 is 1517	37 x 41
The value of words 1, 2, 4, and 6 is 1924	37 x 52
The value of words 1, 2, 4, and 7 is 1813	37 x 49
The value of words 1, 2, 4, 6, and 7 is 2220	37 x 60
The value of words 1, 2, 4, 5, and 6 is 2405	37 x 65
The value of words 1, 2, 3, 4, and 5 is 1998	37 x 54
The value of words 1, 2, 3, 4, 5, and 7 is 2294	37 x 62
The value of words 2, 4, and 5 is 999	37 x 27
The value of words 2, 4, 5, and 7 is 1295	37 x 35
The value of words 2, 4, 5, 6, and 7 is 1702	37 x 46
The value of words 3 and 5 is 481	37 x 13
The value of words 3, 5, and 6 is 888	37 x 24
The value of words 3, 5, 6, and 7 is 1184	37 x 32
The value of word 6 is 407	37 x 11
The value of word 7 is 296	37 x 8
The value of words 6 and 7 is 703	37 x 19
The value of all 7 Hebrew words is 2,701	37 x 73

[16]

The probability that 21 independent occurrences of 37 occurred by chance is 37 to the 21st power or one in 855,531,895,666,462,872,887,391,390,111,637. But that probability pales in comparison to the probability of getting all 21 multiples of 37 in the same sentence that contains more than 16 examples of the number seven.

But now, consider the probability of getting both the features of 7 and 37, as well as the four fundamental components of the universe (time, space, matter, and energy) and the formula for pi (π) in the same seven words. See the next chapter for the formula for pi.

But we still haven't exhausted all the information in this amazing sentence. It also contains the identity of the Creator, the message of salvation, and God's seven-thousand-year prophetic timeline. See the chapter on the Bereshit Prophecy for God's prophetic timeline.

But there is still more. You also have the formula for the fine structure constant (α) in Genesis 1:1, when combined with John 1:1. Refer to the chapter on "The Most Mysterious Number 1/137" for this formula.

If the most powerful artificial general intelligence (AGI) engine in the world could not duplicate all the above features in one seven-word sentence, certainly no human mind could have produced them. The most logical conclusion is that they are the work of the Almighty, the Creator of Time, Space, Matter, and Energy. He left undeniable, irrefutable signposts in a seven-word sentence.

– 7 –

Pi (π) and the Natural Algorithm (e)

What is it that makes pi (π) and the natural algorithm (e) so essential and foundational to our understanding of the universe and the world we live in? Let's begin with pi. This remarkable constant is intricately woven into the fabric of the cosmos, appearing in Einstein's equations that describe the curvature of spacetime, a cornerstone of his theory of general relativity. In quantum physics, pi plays a role in uncertainty principles and quantum wave functions, while in cosmology, it helps to decode cosmic wave patterns and helps us understand the universe's geometry. Pi is essential for calculating the surface areas and volumes of celestial objects like planets, stars, and black holes, and it is indispensable in describing oscillatory motion in quantum mechanics and electromagnetic waves. [17] Simply put, pi is one of the most fundamental and critical constants underlying the structure of the universe.

Now, consider the natural algorithm (e), also known as Euler's number, and its profound role in revealing the fundamental nature and functionality of the universe. This mathematical constant governs dynamic systems, describing exponential processes such as the universe's expansion, the radioactive decay of unstable quantum particles, and the growth of biological systems, neural networks, and spiral galaxies. Additionally, e is crucial in measuring the exponential decay of electric circuits and electromagnetic waveforms. In cosmology, e is featured prominently in Boltzmann's entropy equation, which describes the thermodynamic arrow of time and the directionality of energy flow in the universe. [18]

The relationship between pi and e is also profound, suggesting a deep and perhaps divine connection embedded within mathematics and the cosmos itself. These two constants come together elegantly in Euler's Identity, often celebrated as one of the most beautiful equations in mathematics. Pi and e also feature prominently in the equations of quantum mechanics and statistical mechanics, further underscoring their universal significance.

Now, consider the incredible improbability that both pi and e would appear in the mathematical structure of Genesis 1:1 and John 1:1. Both numbers can be derived to the ten-thousandth decimal place by using the following formula: Multiply the number of letters by the product of the letters and divide by the number of words times the product of the words.

Let's now walk through the formulas for pi and e, beginning with Genesis 1:1.

The number of letters is (28) x the product of the letters is (2.3887872×10^{34}).

The number of words (7) x the product of the words is (3.0415353×10^{17}).

When the numerator is divided by the denominator, the result is 3.1416, or pi, to the 10,000th decimal place. [19]

For a more detailed explanation of this formula, watch the YouTube video: "e and pi are in the Bible - Episode 4: Pi in Genesis 1:1," by Daniel 1210.

The value for e (2.7183) can be found in John 1:1 by using the same formula. Begin by multiplying the number of Greek letters in this verse (52) by the product of those letters, divided by the number of Greek words (17) times the product of those words. The above calculation produces the following results:

The number of letters (52) x the product of the letters is (8.4362514 x 10 to the 73rd power).

The number of words (17) x the product of the words is (9.4930224 x 10 to the 35th power). [20]

When the numerator is divided by the denominator, the result is 2.7183, or e, to the 10,000th decimal place.

Like pi, e is an irrational number, meaning it cannot be expressed as a ratio between two integers. Both are extremely important numbers for science, physics, and engineering calculations. For example, calculus cannot be performed without e.

What is the probability that the formula for pi and e would be embedded in the only two sentences in the Bible that start with the words "In the beginning"? Consider the fact that the first person to accurately calculate pi to the ten thousandth decimal place may have been Zu Chongzhi, sometime between 429 and 500 AD; and Euler's number e wasn't discovered until 1683, by Jacob Bernoulli. [21] Only God, who lives outside of space and time, could have known the importance of these numbers and their significance to understanding our universe before their discovery by Zu Chongzhi and Euler. If God encoded them in His Word hundreds or thousands of years before their discovery, it provides compelling evidence for His existence.

In the chapter titled "The Most Mysterious Number – 1/137," you'll learn that the formula for the fine structure constant (α) also appears in Genesis 1:1 and John 1:1. Consider the fact that these numeric seals, particularly of the numbers 7 and 37, have been undiscovered for thousands of years, embedded within the text of the divinely inspired words of the Old and New Testaments. There are 90,000 possible 5-digit numbers from 1.0000 to 9.9999 that could have occurred from the above equations for pi and e by chance. The probability that the five digits 3.1416 for pi occurred by random chance is one in 90,000. The probability that Euler's number e also occurred by random chance is the same one in 90,000. However, the probability that both occurred by random chance is 1/90,000 x 1/90,000, or 1 chance in 8.1 billion. [22] We haven't factored in the probability of also getting the formula for the fine structure constant α in these same two sentences, which would greatly increase the improbability that they are there by chance.

Another improbable feature of these two verses (Genesis 1:1 and John 1:1) is that the Hebrew and Greek numeric sums of these sentences have factors that are mirror images. For example, the numeric value of the letters in Genesis 1:1 is 2,701 (37 x 73), where 37 is the mirror image of 73. The numeric value of John 1:1 is 3,627 (39 x 93), and 39 is the mirror image of 93. The likelihood that all of the above features examined in these two sentences (Genesis 1:1 and John 1:1) would occur by random chance is far beyond any number that one could imagine.

But we have just scratched the surface regarding the fingerprint, or the Seal of God, in these two divinely inspired Scriptures. There are numerous other hidden features in Genesis 1:1 and John 1:1 that are not covered in this book. For further details on how these two divinely inspired sentences are interconnected, I would encourage you to watch the YouTube video series by Peter Bluer "Bible Numerics Part 1-6."

What we see displayed in the very first sentence of the Bible is a demonstration of complexity and design that staggers the imagination. What else could this be but the fingerprint of God on His instruction book for life? No human author could have devised this many intricate numeric features in a seven-word sentence. And yet, we have still not exhausted the information and hidden numerics in these two Scriptures.

Here is another example: the gematria, or numerical value of the seven words in Genesis 1:1 is (2,701), when added to its reflective number (1,072) it equals 3,773, which also mirrors the factors of 2,701 (37 x 73) and once again shows the number 37 forward and backward. As stated earlier, we see the number 37 forward and backward in the gematria of the Greek word "Logos" (λογοδ), which equals 373.

Genesis 1:1 through Genesis 2:3 is an account of the first seven days of creation. These verses provide additional clues regarding the creator's identity, as these passages have 1,813 Hebrew letters, or (37 x 7 x 7).

Note: The above gematria calculations were provided by Dr. Ivan Panin.

- 8 -

Salvation and the Number 8

God is holy, and He commanded His children in 1 Peter 1:16 (NIV), "Be holy, because I am holy." In Hebrews 12:14 (NIV), we read, "...be holy; without holiness, no one will see the Lord." The Bible also says in Romans 3:23 (NIV), "...all have sinned and fall short of the glory of God." However, the good news is that the Bible says in John 3:16-17 (NIV), "For God so loved the world that he gave his one and only Son, that whoever believes in him shall not perish but have eternal life. For God did not send his Son into the world to condemn the world, but to save the world through him."

Because all have sinned, we need a savior: someone to pay the penalty for our sins and give us new life. We need a savior who can empower us with a new nature—one that can help us overcome our sinful nature. I John 3:4-6 (NIV) says, "Everyone who sins breaks the law; in fact, sin is lawlessness. But you know that he appeared so that he might take away our sins. And in him is no sin. No one who lives in him keeps on sinning. No one who continues to sin has either seen him or known him." He told his disciples in Mark 16:15-16 (NIV), "Go into all the world and preach the gospel to all creation. Whoever believes and is baptized will be saved, but whoever does not believe will be condemned."

In John 3:3 (NIV), Jesus told the Pharisee Nicodemus, "I tell you the truth, unless a man is born again, he cannot see the kingdom of God." Jesus told us that we must be born again, baptized, and filled with the Holy Spirit, for it is the Holy Spirit that empowers us to live a new life. What is incredibly interesting and likely unknown to the writers of the Old and New Testaments is that God underscored these truths using numbers as a divine seal.

The number eight is linked to the salvation message, the names of Jesus, and references to Him. The number eight is believed to signify new beginnings, new birth, resurrection, or salvation. Here are a few examples of how the meaning was derived. God started the world over (a new beginning) with eight people: Noah, his wife, his three sons, and their three wives. God commanded Abraham to circumcise every male child on the eighth day to signify a new covenant between God and man. Genesis 17:12 (KJV) says, "...he that is eight days old shall be circumcised among you, every man child in your generations..." So, circumcision was an outward display of the new covenant between God and Abraham's offspring. God established the seven-day week, and the eighth day begins a new week. Similarly, there are seven notes; the eighth note is the beginning of a new octave. We read in Exodus 22:29b-30 (NIV), "You must give me the firstborn of your sons. Do the same with your cattle and your sheep. Let them stay with their mothers for seven days but give them to me on the eighth day." We read in chapter 14 of the Book of Leviticus that sacrifices for ceremonial cleansings would take place on the eighth day. Jesus is in the lineage of King David, who was the eighth son of Jesse. Jesus was called the Son of David. The birthplace of Jesus (Bethlehem) appears precisely eight times in the New Testament. The above examples are indicators that the number eight is tied to new beginnings, or new life.

In John 3:7 (NIV), Jesus tells Nicodemus, "You should not be surprised at my saying, 'You must be born again.'" Jesus proclaims in John 11:25 (NIV), "I am the resurrection and the life. He who believes in me will live, even though he dies; and whoever lives and believes in me will never die." Jesus is the resurrection, the one who gives new life to believers.

It is not a coincidence that many of the names of Jesus are associated with the number eight. This number represents new beginnings since those who receive Jesus as Lord are born again and raised as new creations in Christ. Let us now look at how this divine truth has been numerically sealed in the gematria values of several of His names.

Messiah has a numeric value of 656	8 x 82
Lord has a numeric value of 800 (Greek)	8 x 100
Christ has a numeric value of 1,480	8 x 37 x 5
Savior has a numeric value of 1,408	8 x 8 x 22

The names mentioned above are associated with the number 8, symbolizing a new beginning or new birth. The following names are connected with both the number 8 and the number 37, the latter representing the Word of God.

Theotes (Godhead) has a value of 592	37 x 8 x 2
Jesus (*Iesous* in Greek) has a value of 888	37 x 8 x 3
Christ has a value of 1,480	37 x 8 x 5
Jesus Christ has a value of 2,386	37 x 8 x 8
Son of Man has a value of 2,960	37 x 8 x 10

[23] [24]

The probability of having five names that are all exactly divisible by both 8 and 37 is (296 x 296 x 296 x 296 x 296), or one chance in 2,272,260,000,000 (2.2 trillion). But hidden within these numbers is another astonishing revelation. If the value for Theotes, θεότης, (592), which refers to the Godhead or Deity, is added to the value of Jesus (888) and the value of Christ (1,480), the result is 2,960, which is the value for "Son of Man." Is this another coincidence or another revelation?

If everyone on Earth had five names, and we assigned values to the letters in their names, you would need more than a trillion people before you would expect to have one person who had all five names exactly divisible by both the numbers 8 and 37 by chance. But that does not account for the numerous other names of Jesus that are also exactly divisible by the numbers 8 or 37, further decreasing the probability that the numerical features of these names could have occurred by chance. Also, consider the fact that many of His names were given thousands of years before

His birth, so they could not have been manipulated to achieve these numeric characteristics during His lifetime.

Here are some additional names or references related to Jesus that are multiples of the number 8 or 37. These references provide further evidence of the astronomical improbability that the numerical values in the gematria of His names or in phrases that refer to Him could have occurred by chance. In describing Himself, Jesus said, "I am the truth," which has a value of 64 (8 x 8). He referred to Himself as the Son of Man 88 times (8 x 11) in the New Testament. The last book in the Bible is the Book of Revelation, which is the revelation of Jesus Christ to John the Apostle. Ivan Panin found that in the original Greek text, there are 888 words (8 x 37 x 3) spoken by Jesus. Again, are these just remarkable coincidences, or were they intentionally ordered so that we might discover them and conclude that they were placed there as a sign or a signature from a divine author?

Another reference to Jesus is found in the Old Testament in Isaiah 9:6 (KJV), which reads, "For unto us a child is born, unto us a son is given: and the government shall be upon his shoulder: and his name shall be called Wonderful, Counselor, the Mighty God, the Everlasting Father, the Prince of Peace." Most Bible scholars believe this Old Testament verse is a direct reference to the Messiah. However, very few are aware that it also contains a hidden code, one that reveals the identity of the person to whom Isaiah was referring.

Since all Scripture is inspired, it is unlikely that Isaiah, writing under the inspiration of the Holy Spirit, had any knowledge of the hidden code embedded in this text. The hidden message that reveals the identity of the Messiah can be found by calculating the gematria of every seventh Hebrew letter in Isaiah 9:6. The Hebrew letters and the values of every seventh letter are shown below in bold font; their sum is 888 (8 x 37 x 3), the same numeric value as Jesus! The probability that this occurred by chance is less than one in a thousand.

Isaiah 9:6

ותהי	לנו	נתר	בר	לנו	ילד	ילר	בי
400		50				30	
ירטצ	פלא	שמו	ויקוא	שבמו	על	שרה	המ
90		6	6			5	
			שלומ	שר	אביער	גבור	אל
			300		1		
30 + 50 + 400 + 5 + 6 + 6 + 90 + 1 + 300 = 888							

Another hidden clue regarding the identity of the Messiah is found in the original Hebrew text of Isaiah 11:1-3a. In English (NIV) it reads, "A shoot will come up from the stump of Jesse; from his roots a Branch will bear fruit. The Spirit of the Lord will rest on him – the Spirit of wisdom and of understanding, the Spirit of counsel and of power, the Spirit of knowledge and of the fear of the Lord – and he will delight in the fear of the Lord." Again, most Bible scholars believe this Scripture refers to the coming Messiah. The individual that Isaiah speaks of has seven spirits. His identity is found in the hidden numeric code embedded within the text. The numeric value (gematria) of every seventh letter in the original Hebrew follows. Once again, the identity of the Messiah is found in the gematria value of every seventh Hebrew letter, which sums to 888 (8 x 37 x 3), the numeric value for Jesus. The probability that the sum of every seventh Hebrew letter in Isaiah 9:6 and Isaiah 11:1-3a would equal 888, both of which are references to the Messiah, is less than one in a million.

Isaiah 11:1-3a

עליו	נגחה	יפרה	משרשיו	ונלר	ישי	מנזע	ויעאחטר
10		5	200		10		200
וגבורה	עצה	רוח	ובינה	חבמה	רוח	יהוה	רוח
3		200		40		6	
יהוה	ביראת	והריחו	יהוה	את	ויר	ראת	רוח
	200	5		1			8
200 + 10 + 200 + 5 + 10 + 6 + 40 + 200 + 3 + 8 + 1 + 5 + 200 = 888							

– 9 –

Sin and the Number 13

We have all heard that the number 13 is supposed to be unlucky. Most hotels and elevators in the United States don't have a 13th floor, and many airlines don't have a 13th row. But few are aware that this number also reinforces the sin message in the Holy Scriptures.

The number 13 in the Bible is often associated with sin and rebellion. For example, in Genesis 14:4 (NIV), we read, "For twelve years they had been subject to Kedorlaomer, but in the thirteenth year they rebelled." In the Book of Joshua, God directed Joshua to lead the Israelites out of Egypt and across the Jordan River into the land of the Amorites and Canaanites, to the city of Jericho. God told him to have the Israelite army march around the city 13 times. On each of the first six days, they marched around the city once. But on the seventh day, they were told to march around the city seven times, totaling 13 trips. When the last trip was completed, the priests blew their trumpets, the people shouted, and the walls of the city came crashing down. The city was destroyed, and all the people were put to the sword, except for Rahab and her family.

One of the reasons God asked Joshua to destroy Jericho and the people of the land that He was giving to the

Israelites was due to their sin in sacrificing children to the god Moloch. 2 Chronicles 28:3 (NIV) says, "He burned sacrifices in the Valley of Ben Hinnom and sacrificed his children in the fire, engaging in the detestable practices of the nations the Lord had driven out before the Israelites."
In 1 Kings 6 and 7, we learn that King Solomon took seven years to build the Lord's Temple but thirteen years to build his own palace.

If the number thirteen represents sin and rebellion, would God not tie this number to the father of sin, to the one who led the rebellion in Heaven? I don't believe it is a coincidence that most of the Hebrew and Greek words associated with Satan have numeric values that are multiples of the number thirteen. In his book, "Number in Scripture, Its Supernatural Design and Spiritual Significance," E.W. Bullinger lists several Hebrew and Greek words, phrases, or verses that refer to Satan. The following examples are from his book.

Satan, in Greek, has a value of 2,197	13 x 13 x 13
Satan, in Hebrew, has a value of 364	13 x 28
Beelzebub (with art.) has a value of 598	13 x 46
Belial has a value of 78	13 x 6
Dragon (Rev 12:9) has a value of 975	13 x 75
Serpent has a value of 780	13 x 60
Murderer has a value of 1820	13 x 140
Tempter has a value of 1053	13 x 81
Fowler has a value of 416	13 x 32

Below are phrases that refer to Satan:

"The power of the enemy" equals 2,509	13 x 193
"That crooked serpent" equals 1,014	13 x 13 x 6
"The power of the air" equals 2,600	13 x 200
"Who is called the Devil and Satan" equals 2,197	13 x 13 x 13
"Your adversary, the devil, as a roaring lion" equals 6,032	13 x 464
"Because the prince of this world is judged" equals 5,577	13 x 13 x 33
"When he speaketh a lie he speaketh of his own, for he is a liar" equals 7,072	13 x 544
"The piercing Serpent, even Leviathan" equals 1,170	13 x 90
"The Dragon that is in the sea" equals 1,469	13 x 113
"They have a king, the angel of the abyss" equals 3,978	13 x 306
"And He said, I beheld Satan as lightning fall from heaven" equals 6,903	13 x 531
"According to the course of this world, according to the prince of the power of the air" equals 9,178	13 x 706
"But against principalities, against powers, against the rulers of the darkness of this world, against spiritual wickedness in high places" equals 1,6211	13 x 1,247

Do you suppose that all the numeric values of these phrases and names of Satan, or Scriptures about him, are multiples of 13 by coincidence? Are all the names of Jesus, which are multiples of 8 and 37, just coincidences? Or did God predetermine these numbers as signposts or markers? Is it possible that God sealed His message regarding sin and rebellion and the way to salvation, in both words and numbers, so that those who are seeking would know the truth?

According to Wikipedia, the numeric values for the letters in the Greek alphabet were assigned in the 6th century BCE, and the Hebrew values, which were adopted from the Greek, were assigned in the 1st century BCE. [25] That's hundreds of years after the books of the Old Testament were written. So, who else but God could have encoded these names and phrases with these numeric seals before the value of the letters had been assigned?

There is a story in the Old Testament that is a foreshadowing of Christ as the sin-bearer, the one who would be hung on a pole and crucified for the sins of the world. We find the story in Numbers 21:4-8 (NIV), where it says, "'They traveled from Mount Hor along the route to the Red Sea, to go around Edom. But the people grew impatient on the way; they spoke against God and against Moses and said, 'Why have you brought us up out of Egypt to die in the desert? There is no bread! There is no water! And we detest this miserable food!' Then the Lord sent venomous snakes among them; they bit the people and many Israelites died. The people came to

Moses and said, 'We sinned when we spoke against the Lord and against you. Pray that the Lord will take the snakes away from us.' So Moses prayed for the people. The Lord said to Moses, 'Make a snake and put it up on a pole; anyone who is bitten can look at it and live.'"

The snake on the pole in the Old Testament was a foreshadowing of Christ's crucifixion in the New Testament. Anyone who was bitten by a poisonous snake could look to the snake on the pole (the representation of sin) and not die. In a similar manner, anyone who is dead in their sins today can put their faith and trust in Jesus Christ, the one who was lifted up on a pole to bear the sins of the world. The Bible says in 1 Pet 2:24 (NIV), "He himself bore our sins in his body on the cross, so that we might die to sins and live for righteousness; by his wounds you have been healed."

Jesus Christ was the sacrificial Lamb of God. Whoever looks upon Him, puts their faith in Him, and submits to His Lordship will receive everlasting life. God told Moses to lift up a snake on a pole to represent the rebellion and sin of the Israelites, so that all who looked at it would be healed. Like the snake that represented the Israelites' sin in Moses' time, Jesus was the representation of our sin as He hung on the cross. All who look to Him have the hope of eternal life.

Here is how Isaiah the prophet foretold the event of how God would lay on him (Jesus) the iniquity of us all, more than 700 years before the event. Isaiah, 53:1-12 (NIV) reads, "Who has believed our message and to whom has the arm of the Lord been revealed? He grew up before him like a tender shoot, and like a root out of dry ground. He had no beauty or majesty to attract us to him, nothing in his appearance that we should desire him. He was despised and rejected by men, a man of sorrows, and familiar with suffering. Like one from whom men hide their faces he was despised, and we esteemed him not. Surely he took up our infirmities and carried our sorrows, yet we considered him stricken by God, smitten by him, and afflicted. But he was pierced for our transgressions, he was crushed for our iniquities; the punishment that brought us peace was upon him, and by his wounds we are healed. We all, like sheep, have gone astray, each of us has turned to our own way; and the Lord has laid on him the iniquity of us all. He was oppressed and afflicted, yet he did not open his mouth; he was led like a lamb to the slaughter, and as a sheep before its shearers is silent, so he did not open his mouth. By oppression and judgment, he was taken away. And who can speak of his descendants? For he was cut off from the land of the living; for the transgression of my people he was stricken. He was assigned a grave with the wicked, and with the rich in his death, though he had done no violence, nor was any deceit in his mouth. Yet it was the Lord's will to crush him and cause him to suffer, and though the Lord makes his life a guilt offering, he will see his offspring and prolong his days, and the will of the Lord will prosper in his hand. After the suffering of his soul, he will see the light of life and be satisfied; by his knowledge my righteous servant will justify many, and he will bear their iniquities. Therefore I will give him a portion among the great, and he will divide the spoils with the strong, because he poured out his life unto death, and was numbered with the transgressors. For he bore the sin of many, and made intercession for the transgressors."

God the Holy Spirit has used the number 13 to reveal numerically what is verbally conveyed in the Scriptures, that Jesus was the sin-bearer, beaten and bruised for our transgressions, and ultimately crucified on a cross. We read the account of Jesus being brought before Pilate for sentencing in Matt 27:19-26 (NIV), "While Pilate was sitting on the judge's seat, his wife sent him this message: Don't have anything to do with that innocent man, for I have suffered a great deal today in a dream because of him. But the chief priests and the elders persuaded the crowd to ask for Barabbas (an insurrectionist and murderer) and to have Jesus executed. Which of the two do you want me to release to you? asked the governor. Barabbas, they answered. What shall I do, then, with Jesus, who is called the Messiah? Pilate asked. They all answered, Crucify him! Why? What crime has he committed? asked Pilate. But they shouted all the louder, Crucify him! When Pilate saw that he was getting nowhere but that instead an uproar was starting, he took water and washed his hands in front of the crowd. I am innocent of this man's blood, he said. It is your responsibility! All the people answered, His blood is on us and on our children! Then he released Barabbas to them. But he had Jesus flogged and handed him over to be crucified."

Bible scholars believe that Jesus received the maximum number of lashes when being flogged, which was forty minus one, or 39. Paul states in II Cor 11:24 (NIV), "Five times I received from the Jews the forty lashes minus one." The significance of 39 lashes is that it represents sin numerically, as 39 is (13 + 13 + 13). Jesus voluntarily took our punishment for sin, represented numerically by the 39 lashes He bore on our behalf.

Some 1,500 years after God told Moses to lift up a snake on a pole for the sins of the people, God's Son was lifted up on a pole to bear the sins of the world. We read in John 19:19-21 (NIV), "Pilate had a notice prepared and fastened to the cross. It read: THIS IS JESUS OF NAZARETH, THE KING OF THE JEWS. The sign was written in Aramaic, Latin, and Greek. The chief priests protested to Pilate, 'Do not write, The King of the Jews,' but that this man claimed to be the king of the Jews." The phrase "Jesus of Nazareth," that hung on the cross above the head of Jesus has a numeric value of 2,197 (13 x 13 x 13). This was God's way of shouting from the heavens, this is my Son, the Lamb of God, who was sacrificed for the sins of the world! What is the probability that Pilate, by coincidence, chose to have those words attached to the cross against the will of the Jewish Pharisees and Sadducees? As 13 is the number for sin, how fitting it is that the words "Jesus of Nazareth," would numerically equal the number for sin to the power of three (13 x 13 x 13)! [26]

But this isn't our only clue that Jesus was the sin-bearer. The sum of the digits for the numeric values of Christ

(1,480) and Savior (1,408) is also 13 (1 + 4 + 8 + 0 = 13) and (1 + 4 + 0 + 8 = 13). Is it a coincidence that Jesus was beaten with (13 + 13 + 13) lashes and that the name "Jesus of Nazareth" was placed on the cross by Pilate, a name that had a numeric value of (13 x 13 x 13)?

But there is an even more amazing illustration of God's love, which is communicated in the gematria, or numerics, of the complete inscription that hung above Jesus on the cross. Each of the four gospels provides a portion of the message. In Matthew 27:37 (NIV), we read, "THIS IS JESUS, THE KING OF THE JEWS." In Mark 15:26 (NIV), it says, "THE KING OF THE JEWS," and in Luke 23:38 (NIV), the words are, "THIS IS THE KING OF THE JEWS." Finally, in John 19:19 (NIV), the inscription reads, "JESUS OF NAZARETH, THE KING OF THE JEWS." When we put the information from all four gospels together, we get the complete inscription, which reads, "THIS IS JESUS OF NAZARETH THE KING OF THE JEWS." Now here is the amazing message of God's love. The number 5 is the number for grace. The value of the Greek letters in the phrase "THIS IS JESUS OF NAZARETH THE KING OF THE JEWS" is 7,215. This number is not only a multiple of 13, indicating that Jesus bore the sin of mankind, but it is amplified in the number for grace, as 7,215 is 13 x 555. [27]

Additionally, we have this example from the Old Testament: the most sacred name for God in Judaism is YHVH, which is translated in the English Bible as "'I AM." God told Moses in Exodus 3:14 (NIV), "I AM WHO I AM. This is what you are to say to the Israelites: 'I AM has sent me to you.'" The gematria value for YHVH is 26 (13 x 2). Jesus told the Pharisees in John 8:58 (NKJV), "Most assuredly, I say to you, before Abraham was, I AM!" As 13 is the number for sin and rebellion, it's not surprising that God also embedded the number 13 in the factors of YHVH, as 26 is (13 x 2), since the "I AM" in Exodus 3:14 is also the sin-bearer.

This revelation is not only found in the factors of the gematria for YHVH; it is also revealed in the Hebrew pictograms for the letters YHVH (Yod, Hey, Vav, Hey). The pictogram for the letter Yod is a nail; the pictogram for Hey is a man with his arms extended upward, and the pictogram for Vav is an arm and hand. So, the message from the pictograms is "Behold the Hand, Behold the Nail." For the meaning of all 22 Hebrew pictograms, reference the chart in the chapter titled "God Declares the End from the Beginning."

In the Book of Romans, we learn that those who willfully sin and refuse to acknowledge God or obey His commandments are given over to a depraved mind. We are also shown what a depraved (or reprobate) mind looks like in Romans 1:21-28 (NIV), which reads, "For although they knew God, they neither glorified him as God nor gave thanks to him, but their thinking became futile, and their foolish hearts were darkened. Although they claimed to be wise, they became fools and exchanged the glory of the immortal God for images made to look like a mortal human being and birds and animals and reptiles. Therefore, God gave them over in the sinful desires of their hearts to sexual impurity for the degrading of their bodies with one another. They exchanged the truth about God for a lie and worshiped and served created things rather than the Creator—who is forever praised. Amen. Because of this, God gave them over to shameful lusts. Even their women exchanged natural sexual relations for unnatural ones. In the same way the men also abandoned natural relations with women and were inflamed with lust for one another. Men committed shameful acts with other men and received in themselves the due penalty for their error. Furthermore, just as they did not think it worthwhile to retain the knowledge of God, so God gave them over to a depraved mind, so that they do what ought not to be done. They have become filled with every kind of wickedness, evil, greed and depravity. They are full of envy, murder, strife, deceit, and malice. They are gossips, slanderers, God-haters, insolent, arrogant and boastful; they invent ways of doing evil; they disobey their parents; they have no understanding, no fidelity, no love, no mercy. Although they know God's righteous decree that those who do such things deserve death, they not only continue to do these very things but also approve of those who practice them."

Jesus came as a sacrificial lamb to take the sins of the world upon Himself. The Creator of the universe left heaven and came to earth. He took the form of a human, with all our frailties, and yet lived a sinless life. He demonstrated His divine nature by healing the sick, opening the eyes of the blind, casting out demons, raising the dead, walking on water, and calming a storm with His words. Then He voluntarily suffered the cruelest of deaths, taking upon Himself the punishment for our sins. He then rose from the dead and appeared to His disciples to show them His resurrected body. He told them He would send the Holy Spirit to be their teacher, guide, and comforter, and just before He was taken into Heaven, He told His disciples, "Go into all the world and preach the good news to all creation. Whoever believes and is baptized will be saved, but whoever does not believe will be condemned." (Mark 16:15-16 NIV)

– 10 –

Wisdom
The Foundation of God's Creation

This chapter examines how the gematria for the Hebrew word for "wisdom" and the creation of the universe are tied to the numbers 37 and 73. The Bible talks about two kinds of wisdom: earthly wisdom, which is man's wisdom, and heavenly wisdom, or God's wisdom. The Hebrew word for wisdom mirrors the numeric phenomena seen in Genesis 1:1, which describes the creation account. The following Bible verses relate "the wisdom of God" to the creation of the heavens and the earth. Psalm 136:5 (KJV) says, "To Him that by wisdom made the heavens." Proverbs 3:19 (NIV) says, "By wisdom the LORD laid the earth's foundations." The book of Jeremiah 10:12 (NIV), says, "... he founded the world by his wisdom and stretched out the heavens by his understanding."

The Hebrew word for wisdom can be calculated using two different gematria approaches. So far, we've only looked at examples of standard gematria, which is derived by summing the assigned values of the Hebrew and Greek letters in a word, a sentence, or a passage of Scripture to obtain its numeric value. Another type of gematria, referred to as ordinal gematria, uses the positions of the letters within the alphabet. In ordinal gematria, the 22 letters in the Hebrew alphabet use sequential values from 1 to 22. The first letter has a value of 1, the second letter has a value of 2, and so on, with each letter taking on the value of its position in the alphabet. The last letter of the Hebrew alphabet has a value of 22. The following table shows both the ordinal and standard gematria values for the Hebrew word for wisdom. Note that the ordinal and standard gematria for the Hebrew word wisdom are the same values found in the factors for the gematria value of Genesis 1:1 (37 x 73). Proverbs 3:19 (NIV) says, "By wisdom the LORD laid the earth's foundations, by understanding He set the heavens in place." We find His signature number, 37, and its reflection, 73, hidden beneath the surface of the text. The author of creation formed the heavens and the earth through wisdom and numerically sealed this fact in the gematria of the Hebrew letters.

5	13	11	8	=	37 (Ordinal Gematria)
ה	מ	כ	ח		(Chokmak translated as Wisdom)
5	40	20	8	=	73 (Standard Gematria)

The ordinal gematria value for the Hebrew word for wisdom is 37. The standard gematria value is 73.

הארץ	ואת	השמימ	את	אלהימ	ברא	בראשית
the earth	and	the heavens	(untranslated)	God	created	In the beginning
296	407	395	401	86	203	913

The standard gematria value for Genesis 1:1 is 2,701, the factors of which are (37 x 73).

Is it just a coincidence that the ordinal and standard gematria values for the Hebrew word for wisdom are 37 and 73? Is it a coincidence that these two numbers are reflective, that both are prime numbers, and that they both have positions within the sequence of prime numbers that are also reflective?

What follows is a list of reflective numbers in the gematria of Genesis 1:1 and John 1:1 and the association of the number 37 with the names of Christ.

- 37 x 73 is the gematria for Genesis 1:1, "In the beginning God created the heavens and the earth."

- 37 is the ordinal gematria, and 73 is the standard gematria for the Hebrew word for wisdom, by which God made the heavens and earth.

- 39 x 93 is the gematria for John 1:1, "In the beginning was the Word, and the Word was with God, and the Word was God."

- 37 and 73 are the 12th and 21st prime numbers, and 12 and 21 are reflective numbers.

- 12 squared is 144, and 21 squared is 441; 144 and 441 are reflective numbers.

- The factors of 441 are (3 x 7 x 7 x 3), and again, we see a reflection in the numbers 3 and 7 and 7 and 3.

- The 37th odd number is 73, and these numbers are reflective.

- 37 is a factor in the gematria for the Greek names: Jesus 888 (37 x 8 x 3), Christ 1480 (37 x 8 x 5), Jesus Christ 2,368 (37 x 8 x 8), Theotes 592 (37 x 8 x 2), and Son of Man 2,960 (37 x 8 x 10).

- 373 is the gematria for Logos or Word; 37 is seen both forward and backward in this number.

Hebrews 1:3 (NIV) says, "The Son is the radiance of God's glory and the exact representation of His being, sustaining all things by His powerful word. After He had provided purification for sins, He sat down at the right hand of the Majesty in heaven." Both Genesis 1:1 and John 1:1 have factors that are reflective (37 x 73) and (39 x 93), and both also contain references to the Creator (Jesus), who is the exact representation (or reflection) of God the Father, according to the Bible. In John 1:1, Jesus appears as "the Word," and in Genesis 1:1, He appears as the "Alpha and Omega," or in Hebrew, the "Aleph and Tav."

– 11 –

God's Seal in the Human Genome

If God put a divine numeric seal on the Bible, is it possible that He also put His divine seal on life itself, in the genetic code? Let us again examine the numbers 8 and 37, which are encoded in the names of Jesus: 888 (37 x 8 x 3), Christ: 1,480 (37 x 8 x 5), Jesus Christ: 2,368 (37 x 8 x 8), Theotes: 592 (37 x 8 x 2), Son of Man: 2,960 (37 x 8 x 10). The Bible says in John 1:3 (NIV), "Through him all things were made." If this is true, is it surprising that scientists and genealogists have recently discovered a pattern of 8s and 37s in the genetic source code for all life forms, from bacteria to humans?

DNA, or deoxyribonucleic acid, carries the blueprint instructions for the development, growth, reproduction, and internal functions of every cell in living organisms. DNA is composed of four nucleotides: cytosine (C), guanine (G), adenine (A), and thymine (T). "The nuclear genome comprises approximately 3,200,000,000 nucleotides of DNA, divided into 24 linear molecules, the shortest being 50,000,000 nucleotides in length and the longest 260,000,000 nucleotides..." [28] Here again, we see the number eight in the factors of the 24 (8 + 8 + 8) linear molecules. In his book *Signature in the Cell, DNA and the Evidence for Intelligent Design*, Dr. Stephen C. Meyer states, "This string of alphabetic characters looks as if it could be a block of encoded information, perhaps a section of text or machine code. That impression is entirely correct, for this string of characters is not just a random assortment of the four letters A, T, G, and C, but a representation of part of the sequence of genetic assembly instructions for building a protein machine – an RNA polymerase – critical to gene expression (or information processing) in a living cell." [29, p. 23]

The average human body contains approximately 32 trillion cells. [30] "There are as many as 100,000 unique types of proteins within a typical human cell." [31] These proteins carry out a multitude of functions within the cell. The self-replicating DNA instructions for building every part of our being are embedded in the cells that make up the human body. Not only are these DNA instructions self-replicating, but they also produce enzymes that read and correct errors in the DNA molecules being copied. The level of complexity underlying these self-replicating, self-correcting instruction processes embedded in DNA far surpasses the complexity of anything mankind has ever produced or invented. In fact, even with all our recent discoveries in mapping the human genome, much of the cell's inner workings are still beyond our understanding.

Another characteristic that makes the information encoded in DNA absolutely mind-boggling is the fact that it is multi-layered. Unlike computer code, which is read linearly, DNA code has different instructions depending on where the program begins reading. In addition, the chromosomes fold and connect to other chromosomes within the DNA helix, which results in different information being processed. There is also a time dimension for processing DNA information. But the most astounding characteristic of DNA may be its capacity for storing information. Scientists have discovered that DNA has the most densely compacted storage capacity of anything known to man.

Due to its enormous storage capacity, DNA is now being promoted as the most effective way of storing the world's ever-growing volume of data. "Consider this: humanity will generate an estimated 33 zettabytes of data by 2025— that's 3.3 followed by 22 zeroes. DNA storage can squeeze all that information into a storage device the size of a ping-pong ball, with room to spare. The 74 million million bytes of information in the Library of Congress could be crammed into a DNA archive the size of a poppy seed— 6,000 times over." [32] The superiority of DNA as a data storage medium provides strong evidence for its supernatural creation.

The following table shows the specific amino acids or DNA base triplets that make up the standard genetic code. Note that there are 64 (8 x 8) DNA base triplets.

		SECOND LETTER					
		U	C	A	G		
F I R S T L E T T E R	U	UUU - Phe UUC - Phe UUA - Leu UUG - Leu	UCU - Ser UCC - Ser UCA - Ser UCG - Ser	UAU - Tyr UAC - Tyr UAA Stop UAG Stop	UGU - Cys UGC - Cys UGA Stop UGG Trp	U C A G	T H I R D L E T T E R
	C	CUU - Leu CUC - Leu CUA - Leu CUG - Leu	CCU - Pro CCC - Pro CCA - Pro CCG - Pro	CAU - His CAC - His CAA - Gln CAG - Gln	CGU - Arg CGC - Arg CGA - Arg CGG - Arg	U C A G	
	A	AUU - Ile AUC - Ile AUA - Ile AUG Met	ACU - Thr ACC - Thr ACA - Thr ACG - Thr	AAU - Asn AAC - Asn AAA - Lys AAG - Lys	AGU - Ser AGC - Ser AGA - Arg AGG - Arg	U C A G	
	G	GUU - Val GUC - Val GUA - Val GUG - Val	GCU - Ala GCC - Ala GCA - Ala GCG - Ala	GAU - Asp GAC - Asp GAA - Glu GAG - Glu	GGU - Gly GGC - Gly GGA - Gly GGG - Gly	U C A G	

Image credit: "The genetic code," by OpenStax College, Biology (CC BY 3.0). This figure shows the genetic code for translating each nucleotide triplet in mRNA into an amino acid or a termination signal in a protein. (credit: modification of work by NIH). [33]

Each three-letter sequence of mRNA nucleotides corresponds to a specific amino acid or to a stop codon. UGA, UAA, and UAG are the stop codons. AUG is the codon for methionine and is also the start codon. Dr. Meyer noted in his book *Signature in the Cell*, "What humans recognize as information...originates from thought – from conscience or intelligent activity." He states, "The gene is a package of information, not an object. The pattern of base pairs in a DNA molecule specifies the gene. But the DNA molecule is the medium, it is not the message." [34] He then asks the question, "What does it mean when we find information in natural objects – living cells – that we did not ourselves design or create? The genetic code, found in all living organisms, is the set of rules that translate DNA into proteins (the building blocks of all living cells) and does not alter as it is passed down through generations. The building blocks of proteins are amino acids, and thus, the amino acids represent the foundational building blocks of life." [35]

All life forms have 20 universal amino acids that directly code for genes. In a paper entitled; "The WOW! signal of the terrestrial genetic code", Vladimir I. Shcherbak and Maxim A. Makukov demonstrate how the nucleons in the human genome are mathematically arranged according to each of the triple repdigits (111 through 999). [36] This may be another incredible clue as to the author of the human genome because each of the triple repdigits contains a number that is exactly divisible by 37, the number we learned earlier represents the Word of God.

111	(1 + 1 + 1 = 3 and 1 x 3 x 37 = 111)
222	(2 + 2 + 2 = 6 and 2 x 3 x 37 = 222)
333	(3 + 3 + 3 = 9 and 3 x 3 x 37 = 333)
444	(4 + 4 + 4 = 12 and 4 x 3 x 37 = 444)
555	(5 + 5 + 5 = 15 and 5 x 3 x 37 = 555)
666	(6 + 6 + 6 = 18 and 6 x 3 x 37 = 666)
777	(7 + 7 + 7 = 21 and 7 x 3 x 37 = 777)
888	(8 + 8 + 8 = 24 and 8 x 3 x 37 = 888)
999	(9 + 9 + 9 = 27 and 9 x 3 x 37 = 999)

If we sum the numbers 111 through 999, we get the number 4,995 (37 x 135). Coincidentally, this is also the same numerical value as the sum of the values of the 27 letters in the Greek alphabet used to write the New Testament! [37] Coincidence or another signpost?

The authors drew the following conclusion: "Accurate and systematic, these underlying patterns appear as a product of precision logic and nontrivial computing rather than of stochastic processes (the null hypothesis that they are due to chance coupled with presumable evolutionary pathways is rejected with P-value <10-13)." In other words, the authors reject the idea that this could have happened by chance through evolution, calculating the probability to be less than one chance in ten trillion. [38]

In their research paper, published in Icarus, Shcherbak and Makukov noted that the number 37 also appears in the molecular mass of the 20 amino acids, the building blocks from which all living organisms are constructed. For example:

The total molecular mass of the 20 amino acids = 2,738 (37 x 74), and 74 is the 37th even number, and its factors are (37 x 2).

The total mass of the molecular core shared by all 20 amino acids is 74 (37 x 2), and 74 is the 37th even number.

Of the 20 amino acids, 19 have 73 (the reflection of 37) nucleons (protons and neutrons).

Proline, the only amino acid out of the 20 that is different, has 74 nucleons (37 x 2) and 74 is the 37th even number.

Mitochondrial DNA contains exactly 37 genes, all of which are essential for normal mitochondrial function.

There are 28 codons (3 structural units within DNA), each with a total atomic mass of 1,665 (37 x 45) and a combined side chain atomic mass of 703 (37 x 19); and 703 is the 37th triangular number. [39]

The European Organization for Nuclear Research, known as CERN (Conseil Européen pour la Recherche Nucléaire), located in Geneva, Switzerland, is the site of the world's largest and highest-energy particle collider. In an interview between Ken Campbell and the former Director of CERN, they discussed the subject of quantum physics and quantum electrical fields. The Director of CERN asked Ken Campbell how many different quantum electrical fields he thought were necessary to build everything we see in the known universe. Ken Campbell assumed that it must be countless billions. The Director of CERN replied, "No, no, 37 different fields; that is the minimum number of fields that we need to explain all that we know about particle physics. It appears that everything in the known universe – electrons, quarks, and gluons, etc. – has been created from 37 blueprints called quantum electrical fields." [40]

Once again, we see God's signature imprint, the number 37, in the number of quantum electrical fields necessary for the creation of all matter. To suggest that the number 37 appears in Genesis 1:1, in the names of Jesus, in the genetic code for life, and in quantum electrical fields is just a coincidence is to ignore God's fingerprint, prominently displayed in both His Word and in His creation. This signature number, 37, points to the author of the Bible and to the Creator of the universe as the same divine being. He has left unmistakable clues of His existence for all to see. Romans 1:20 (NIV) says, "For since the creation of the world God's invisible qualities – his eternal power and divine nature – have been clearly seen, being understood from what has been made, so that men are without excuse."

Today's generation has more scientific information than any other in history. Consequently, we have more insight into the detailed inner workings of God's creation and the enormous complexity found in all life forms. We now know that there are no simple life forms, like those envisioned by Charles Darwin. As Michael Denton, PhD, and Senior Fellow at the Discovery Institute's Center for Science and Culture, states, "To grasp the reality of life as it has been revealed by molecular biology, we must magnify a cell a thousand million times until it is 20 kilometers in diameter and resembles a giant airship large enough to cover a great city like London or New York. What we would then see would be an object of unparalleled complexity and adaptive design. On the surface of the cell, we would see millions of openings, like the portholes of a vast spaceship, opening and closing to allow a continual stream of materials to flow in and out. If we were to enter one of these openings, we would find ourselves in a world of supreme technology and bewildering complexity." [41]

Modern science has now mapped the human genome and provided detailed descriptions of the enormous information storage capacity of DNA molecules. It has given us a greater understanding of the extreme complexity of even the simplest cells, which are many orders of magnitude greater than anything Charles Darwin could have imagined. Today, we are much further away from being able to explain how life could have spontaneously generated from inorganic material than we were in the 1800s.

We have reached a point where the evidence is overwhelming that life could not have arisen from a primordial soup. The primordial Earth had no DNA molecules, no RNA molecules, no protein molecules, and no instructions for building the complex machinery necessary to produce and sustain life. Since all the life-sustaining building blocks must be simultaneously present for life to exist, there remains no reasonable explanation for how life could have spontaneously come into existence. Recent advances in our understanding of the complexity and inner workings of cells have not brought us any closer to producing a living cell in the laboratory. Even if given all the individual components necessary to make a cell and pristine laboratory conditions, we are unable to make even the simplest of cells. The more we learn, the harder it gets!

- 12 -

The Most Mysterious Number 1/137

The number 1/137, which represents the fine structure constant, is possibly the most mysterious number ever discovered by modern science. According to astrophysicist Paul Davies, a professor at Arizona State University and Director of BEYOND, the Center for Fundamental Concepts in Science, this mysterious number appears at the intersection of relativity, electromagnetism, and quantum mechanics. [42] Paul Dirac, a theoretical physicist and one of the founders of quantum mechanics and quantum electrodynamics, said that the fine structure constant is "the most fundamental unsolved problem in physics." [43] Eric Cornell, a Nobel Prize-winning physicist at the University of Colorado and the National Institute of Standards and Technology, said, "In the physics of low-energy matter—atoms, molecules, chemistry, biology—there's always a ratio of bigger things to smaller things. Those ratios tend to be powers of the fine-structure constant." [44]

The fine structure constant 1/137 is represented by the Greek symbol alpha (α) and is considered the best example of a pure number, meaning that it doesn't require units of measure, such as grams, pounds, kilograms, meters, miles, seconds, minutes, hours, etc. It is a non-dimensional atomic constant used to measure the probability of electrons or other charged particles emitting or absorbing photons. It also measures the ratio of the strength of electromagnetism to the strong nuclear force. Some have suggested that the fine structure constant may be the key to everything we see around us. It combines three other fundamental constants: the speed of light, the charge carried by a single electron, and Planck's constant, the latter being a universal constant that defines the nature of energy on a quantum level and measures the frequency and energy produced by photons. In an article titled "Life as we know it would not exist without this highly unusual number," the author Paul Sutter stated, "If it had any other value, life as we know it would be impossible. And yet we have no idea where it comes from." [45]

What makes this number 1/137 so interesting and perplexing is that no one knows where the number comes from or why it seems to be foundational to the structure of the physical universe. It shows up everywhere and has yet to be explained. In his April 22, 2011 article, "How Many Fundamental Constants Are There?" John Baez stated, "The observed values of the dimensionless physical constants (such as the fine-structure constant) governing the four fundamental interactions [gravity, electromagnetism, the strong nuclear force, and weak nuclear force] are balanced as if fine-tuned to permit the formation of commonly found matter and subsequently the emergence of life." [46] This mysterious number also appears to be tied to the laws that govern the structure and behavior of the universe, such as general relativity, quantum physics, subatomic particle movements, the force of gravity, and electromagnetism.

According to some of the world's brightest physicists, the fine structure constant 1/137, or alpha, is the single most revealing message that Earth could send to an alien planet to convince them that we have a good understanding of the fundamental science and the physics of the universe. It's also interesting that the numerical value of 1/137 is equal to .007 when rounded to the nearest thousandth. Seven, you'll recall, is the number most associated with God the Father, and it means completeness, perfection, or holiness. [47]

Once again, God has provided a remarkable numeric clue regarding the origin of this mysterious number 1/137. He encoded it in Genesis 1:1 and John 1:1, the only two sentences in the Bible that reference "In the Beginning." The digits of α, the fine structure constant, can be derived with 99.999% accuracy by joining the gematria values of Genesis 1:1 (2701) and John 1:1 (3627) and squaring them (27013627^2). The result is 7.2973e14. The value of α, or the fine structure constant, is 0.0072973... x 10e-3. [48]

Given that the number 137, or 1/137, is the most mysterious and perplexing number in quantum physics and that it appears to be foundational to the structure of the entire universe, is it surprising to learn that the Creator of the universe also left us with another clue to His existence in the first seven digits of pi (π)? The value associated with the universal constant, α, is 137, and it can be derived by squaring the first seven digits of pi and then adding them together.

$\pi = 3.141592$ or, $3^2 + 1^2 + 4^2 + 1^2 + 5^2 + 9^2 + 2^2 = 137$ [49]

God's use of numeric signatures in Scripture, in biology, and in values for cosmological constants like the fine structure constant may be His way of proclaiming His existence with numeric clues. In this mysterious number (137), we see several other numbers that may provide further clues to the identity of our Creator. The first number we see is 1. There is one God (the Father) over everything. We also see the number 7, both in the decimal equivalent of 1/137 (.007) and in the last digit of 137. We have learned that 7 is the number most frequently associated with God the Father and is also a number representing perfection. We also see the number 37 in the last two digits. This number represents the Word of God, Jesus Christ, the second person in the Trinity. Finally, we see the number 3, which could represent the Holy Trinity: the Father, Son, and Holy Spirit.

Could it be that the mystery behind the fine structure constant (1/137) and its decimal equivalent (0.007), or (.0073) if extended to the fourth decimal place, is further evidence that will ultimately lead us to the author of creation? Is it possible that the fine structure constant is just one more sign that points to the Creator? Are we to believe that all these numbers are occurring by mere coincidence? Is it possible that God has left us with an abundance of clues to His identity so that we are without excuse?

Of all the symbols that could have been chosen to represent the fine structure constant, the one chosen was the Greek letter α, alpha. The letter alpha, or aleph in Hebrew, is symbolic of the Godhead or Leader. Its numeric value is 1, as in the one true God.

Describing Himself to the apostle John in Revelation 21:6b (NIV), Jesus said, "I am the Alpha and Omega, the Beginning and the End." The Greek letters Alpha and Omega are the alphabetic equivalents of the Hebrew letters Aleph and Tav. These two letters appear together in the middle of the first sentence in the Bible, Genesis 1:1. There is no English equivalent word for these two letters, they are untranslated. What we may be seeing embedded in the center of the creation sentence is a depiction of the Creator, the Alpha and Omega, or the Aleph and Tav.

These two letters, Aleph and Tav, appear 7,339 times in the Old Testament and are always untranslated in our English Bible. Is it possible we have 7,339 Old Testament clues regarding the identity of the Messiah? Could it be that the Christ of the New Testament, the Alpha and Omega of the Book of Revelation, is referring to Himself in the Old Testament as the Aleph and Tav?

Given all we know, is it not likely, even highly likely, that the God of the Bible embedded within the Holy Scriptures two Hebrew letters that point to the identity of His Son, the one who referred to Himself as the Alpha and Omega, the First and the Last? Is it coincidence or divine revelation that the pictogram meanings for the letters Aleph and Tav are Godhead and Covenant and that the symbols representing them are an ox, which looks like an A on its side, and a cross? Jesus referred to Himself as the Alpha and Omega, but speaking in Aramaic, He would have referred to Himself as the Aleph and the Tav.

– 13 –

Why are Fibonacci Numbers Everywhere?

Leonardo da Pisa, also known as Leonardo Fibonacci, was an Italian mathematician. He is best known for his book, "Liber Abaci," which introduced Europe to Arabic numbers in 1202. [50] He is also known for introducing us to a sequence of numbers known as the Fibonacci sequence, or the Fibonacci numbers. Fibonacci numbers are formed by starting with 0 and 1 and then adding them together to get the next number in the sequence. Each subsequent number is derived by adding the two previous numbers in the sequence. For example: 0 + 1 = 1, 1 + 1 = 2, 1 + 2 = 3, 2 + 3 = 5, 3 + 5 = 8, 5 + 8 = 13, and so on.

The Fibonacci formula produces the following sequence of numbers, which we refer to as Fibonacci numbers: 0, 1, 2, 3, 5, 8, 13, 21, 34, 55, 89, 144, 233… and the sequence continues indefinitely. What makes these numbers unique is the peculiar formula or method used to derive the next number in the sequence. They are not random but rather follow a distinct pattern. When a Fibonacci number is divided by the previous number in the sequence, it produces a ratio that approaches a value of 1.618; for example, 89 divided by 55 equals 1.618. This ratio was initially referred to as the "Divine Proportion" and later became known as the golden ratio. It is represented by the Greek letter Phi.

Fibonacci numbers can be seen everywhere in nature, from quantum mechanics to the structure of our universe. We see them in spider webs, pinecones, pineapples, flowers, nautilus shells, hurricane spirals, galaxy spirals, and even the curvature of ocean waves. How did nature, or random chance, create such an unusual and unique sequence of numbers, and why do these numbers produce optimal and efficient growth patterns wherever we see them? [51] Optimization and efficiency are thought to be byproducts of intelligence. So why is it that when Fibonacci numbers are observed in nature, they are not recognized as the product of an intelligent designer?

Should we believe that nature, by random chance, created this unusual numbering sequence, a sequence that optimizes efficiency? Is it not more likely that an intelligent designer created this sequence, knowing that intelligent beings would be able to discover it and conclude that there must have been an intelligent mind behind the pattern? Since the process of optimization requires intelligence, is it more likely the optimization we see in nature occurred by random chance, or is it more likely that an intelligence far greater than our own was responsible?

Let's look at some examples of Fibonacci numbers that appear in nature, starting with flowers. The number of petals on flowers is likely to be a Fibonacci number: 3, 5, 8, 13, 21, 34, and 55. If you look at the florets on a sunflower, you will find that they are arranged in spirals of 21, 34, and 55, all of which are Fibonacci numbers. The fruitlets on a pineapple and pinecone are arranged using Fibonacci numbers. If you look at the seeds in the spiral of a sunflower, you'll find that the seeds are laid out in a spiral pattern of 1.618 turns per seed (the golden ratio) or every 137.5 degrees. This number, 137.5, is referred to as the golden angle, and it also appears throughout nature.

Where have we seen that number 137? You'll recall that it is the number connected to the fine structure constant, the most mysterious and possibly the most important number in the universe. It appears to be foundational to the very existence of our universe and is represented by the symbol alpha (α). It is found at the intersection of relativity, electromagnetism, and quantum mechanics, and it is considered the best example of a pure number. We now learn that it also appears in the golden angle, which is tied to the Fibonacci ratio, sometimes referred to as the "Divine Proportion." Are we to believe that this is just another coincidence? Or is it much more likely to be a sign, a fingerprint, or the signature of the one who made everything?

In a PubMed article from the NIH National Library of Medicine titled "Fibonacci Sequence and Supramolecular Structure of DNA," the author states: "...[the] 3D structure of DNA molecule is assembled in accordance with a mathematic rule known as the Fibonacci sequence." [52] In another PubMed article dated April 2008 titled, "Nucleotide frequencies in the human genome and

Fibonacci numbers," the following statements appear: "This result may be used as evidence for the Fibonacci string model that was proposed to explain the sequence growth of DNA repetitive sequences. Notably, the predicted values are solutions to an optimization problem, which is commonplace in many of nature's phenomena." [53]

What are these articles suggesting? Is it that the predicted values (the Fibonacci numbers) are solutions to an optimization problem? How does nonintelligent matter develop highly complex optimization solutions using Fibonacci numbers? Wherever we find solutions to complex problems, we find an intelligent mind at work.

All DNA molecules are made up of two intertwined double-helix spirals. Is it surprising to learn that we find Fibonacci numbers underlying these spirals, as each full cycle measures 21 angstroms in width and 34 angstroms in length? [54] Fibonacci numbers, or the golden ratio, or the divine proportion, are also found in the most revered item in the Old Testament, the Ark of the Covenant. We read in Exodus 25:10 (KJV), "And they shall make an ark of shittim wood: two cubits and a half shall be the length thereof, a cubit and a half the breadth thereof, and a cubit and a half the height thereof." The ratios between the height, width, and length of the ark are Fibonacci numbers, expressed in half cubits: 3 and 5.

A mathematics lesson for 4th-grade elementary students in the state of Hawaii was titled, "The Fibonacci Sequence in Nature – Are patterns consistent in nature, and what is the connection between patterns in nature and the Fibonacci sequence?" The lesson aimed to teach students how to recognize standard benchmarks and values and analyze patterns. A short passage from the lesson follows: "Nature is all about math. If you were to observe the way a plant grows new leaves, stems, and petals, you would notice that it grows in a pattern following the Fibonacci sequence. Plants do not realize that their growth follows this sequence. Rather, plants grow in the most efficient way possible – new leaves and petals naturally grow in spaces between old leaves, but there is always enough room left for one more leaf or petal to grow." [55]

The explanation given for how Fibonacci numbers appear in plants to optimize the positioning of stems, new leaves, and petals is that "plants grow in the most efficient way possible." [56] The lesson rightly acknowledges that plants do not realize their growth follows the Fibonacci sequence. However, the extreme intelligence behind these optimization patterns that we find throughout nature is obvious. If the intelligence that optimizes for efficient growth patterns in plants does not lie within the plant, then where does it come from? There are only two choices. It either comes from an omnipotent Creator who left His signature in the form of Fibonacci numbers, or it is the result of random chance and natural selection, neither of which has intellectual ability. Which is more plausible?

For additional information on where Fibonacci numbers are found in nature, I recommend viewing the YouTube video "God's Fingerprint – The Fibonacci Sequence – Golden Ratio and the Fractal Nature of Reality." [57]

– 14 –

The Unspeakable Name

In Judaism, the most revered and holy name of God is YHWH. The name is considered too sacred to be spoken aloud in some Jewish communities. The name is translated as "I AM" in English and consists of four Hebrew letters: Yod (י), Hey (ה), Vav (ו), and Hey (ה). In Proverbs 25:2 (NIV), we read, "It is the glory of God to conceal a matter; to search out a matter is the glory of kings." Hidden within this unspeakable name of God lies a profound revelation—the identity of the Messiah.

The first clue comes from the Hebrew pictograms that correspond to the letters of YHWH. Hebrew letters not only carry numeric values but are also represented by pictograms that provide additional layers of meaning. The letter Yod, with a numeric value of 10, is symbolized by a hand or an arm and hand. The letter Hey, valued at 5, is depicted as a man with raised arms or a window, signifying "look" or "behold." The letter Vav, with a value of 6, represents a nail or tent peg. Together, these symbols paint a striking picture. Combining them all together yields the message: "Behold the hand, behold the nail!" This imagery foreshadows the crucifixion of Jesus, the Son of God, who was sacrificed as the Lamb of God for the sins of the world on the Feast of Passover.

The second clue lies in the gematria—the Hebrew numeric value—of YHWH, which totals 26 (10 + 5 + 6 + 5). Remarkably, the number 8, frequently associated with Jesus through names like Christ, Savior, and Son of Man, is embedded within the value, YHWH, as the digits 2 and 6 sum to 8. Additionally, the factors of 26 (13 x 2) reveal the number 13, symbolizing sin or rebellion. This connection underscores Jesus as the sin-bearer, the one who bore the punishment for humanity's sin and rebellion.

Similar numeric patterns of 8, the number for new beginnings, and 13, the number for sin and rebellion, are also found in other names by which Jesus was known. For example, Christ has a gematria of 1,408, and the digits 1 + 4 + 0 + 8 = 13, and its factors are (37 x 8 x 5). Savior has a gematria of 1,048, and the digits 1 + 0 + 4 + 8 = 13, and its factors are (8 x 131). Jesus of Nazareth has a gematria of 2,197 (13 x 13 x 13), and its factors are (37 x 8 x 10). The number 8 points to Christ as the Savior, and the number 13 points to Him as the sacrificial sin offering. As Hebrews 10:11-14 states: "But this Man (Jesus), after He had offered one sacrifice for sins forever, sat down at the right hand of God...For by one offering, He has perfected forever those who are being sanctified."

According to Rabbi Michael Skobac, the three letters used in the name YHWH permute into three words whose meanings are: He was, He is, and He will be; that is, the Eternal God. [58] This eternal identity is echoed in Revelation 1:4 (NIV), where Jesus appears to John after

His resurrection and proclaims: "Grace and peace to you from him who is, and who was, and who is to come..."

From Genesis to Revelation, God reveals the identity of the Savior. While this truth has been spiritually concealed, the veil is being lifted in these last days, as prophesied by Paul in Romans 11:25-27 (NIV): "Israel has experienced a hardening in part until the full number of the Gentiles has come in. And so all Israel will be saved, as it is written: 'The deliverer will come from Zion; he will turn godlessness away from Jacob. And this is my covenant with them when I take away their sins.'" Likewise, we read in Daniel 12:1 (NIV), "At that time Michael, the great prince who protects your people, will arise. There will be a time of distress such as has not happened from the beginning of nations until then. But at that time your people—everyone whose name is found written in the book—will be delivered."

Through divine numeric patterns, pictograms, and the associated meanings of the numbers in His names, God has revealed His Son, Jesus, as the Messiah, the Savior, and the sin-bearer, fulfilling His divine plan of redemption.

- 15 -

Israel's Messiah

What does the Old Testament say regarding the Messiah? Are there any clear references in the Bible that foretell His coming, His mission, and His crucifixion? The answer is an emphatic YES! Probably the most famous and clearest text can be found in chapter 53 of the Book of Isaiah.

Here is the text from Isaiah 53:1-12 (NIV), "Who has believed our message and to whom has the arm of the Lord been revealed? He grew up before him like a tender shoot, and like a root out of dry ground. He had no beauty or majesty to attract us to him, nothing in his appearance that we should desire him. He was despised and rejected by men, a man of sorrows, and familiar with suffering. Like one from whom people hide their faces he was despised, and we esteemed him not. Surely he took up our infirmities and carried our sorrows, yet we considered him stricken by God, smitten by him, and afflicted. But he was pierced for our transgressions, he was crushed for our iniquities; the punishment that brought us peace was on him, and by his wounds we are healed. We all, like sheep, have gone astray, each of us has turned to our own way; and the Lord has laid on him the iniquity of us all. He was oppressed and afflicted, yet he did not open his mouth; he was led like a lamb to the slaughter, and as a sheep before its shearers is silent, so he did not open his mouth. By oppression and judgment, he was taken away. And who can speak of his descendants? For he was cut off from the land of the living; for the transgression of my people, he was stricken. He was assigned a grave with the wicked, and with the rich in his death, though he had done no violence, nor was any deceit in his mouth. Yet it was the Lord's will to crush him and cause him to suffer, and though the Lord makes his life a guilt offering for sin, he will see his offspring and prolong his days, and the will of the Lord will prosper in his hand. After the suffering of his soul, he will see the light of life and be satisfied; by his knowledge my righteous servant will justify many, and he will bear their iniquities. Therefore, I will give him a portion among the great, and he will divide the spoils with the strong, because he poured out his life unto death, and was numbered with the transgressors. For he bore the sin of many and made intercession for the transgressors."

Another passage from the Book of Psalms provides an equally compelling prophecy regarding the coming Jewish Messiah. The text is found in Psalm 22:1-31 (NIV), and reads, "My God, my God, why have you forsaken me? Why are you so far from saving me, so far from the words of my groaning? O my God, I cry out by day, but you do not answer, by night, and am not silent. Yet you are enthroned as the Holy One; you are the praise of Israel. In you our fathers put their trust; they trusted and you delivered them. They cried to you and were saved; in you they trusted and were not disappointed. But I am a worm and not a man, scorned by men and despised by the people. All who see me mock me; they hurl insults, shaking their heads. 'He trusts in the Lord; let the Lord rescue him. Let him deliver him, since he delights in him.' Yet you brought me out of the womb; you made me trust in you, even at my mother's breast. From birth I was cast upon you; from my mother's womb you have been my God. Do not be far from me, for trouble is near and there is no one to help. Many bulls surround me; strong bulls of Bashan encircle me. Roaring lions that tear their prey open their mouths wide against me. I am poured out like water, and all my bones are out of joint. My heart has turned to wax; it has melted within me. My strength is dried up like a potsherd, and my tongue sticks to the roof of my mouth; you lay me in the dust of death. Dogs have surrounded me; a band of evil men has encircled me; they have pierced my hands and my feet. I can count all my bones; people stare and gloat over me. They divide my garments among them and cast lots for my clothing. But you, O Lord, be not far off; O my Strength, come quickly to help me. Deliver my life from the sword, my precious life from the power of the dogs. Rescue me from the mouth of the lions; save me from the horns of the wild oxen. I will declare your name to my brothers; in the congregation I will praise you. You who fear the Lord, praise him! All you descendants of Jacob, honor him! Revere him, all you descendants of Israel! For he has not despised or disdained the suffering of the afflicted one; he has not hidden his face from him but has listened to his cry for help. From you comes my praise in the great assembly; before those who fear you will I fulfill my vows. The poor will eat and be satisfied; they who seek the Lord will praise him – may your hearts live forever! All the ends of the earth will remember and turn to the Lord, and all the families of the nations will bow down before him for dominion belongs to the Lord and he rules over the nations. All the rich of the earth will feast and worship; all who go down to the dust will kneel before him – those who cannot keep themselves alive. Posterity will serve him; future generations will be told about the Lord. They will proclaim his righteousness to a people yet unborn – for he has done it."

Psalm 69 provides additional references foretelling how the nation of Israel would reject and scorn Him. Psalm 69:7-10 (NIV) says, "For I endure scorn for your sake, and shame covers my face. I am a stranger to my brothers, and alien to my own mother's sons; for zeal for your house consumes me, and the insults of those who insult you fall on me. When I weep and fast, I must endure scorn..." Later in the same chapter, we read in Psalm 69:20-21 (NIV), "Scorn has broken my heart and has left me helpless; I looked for sympathy, but there was none, for comforters, but I found none. They put gall in my food and gave me vinegar for my thirst."

Regarding the coming of Zion's King, it says in Zechariah 9:9 (NIV), "Rejoice greatly, O Daughter of Zion! Shout, Daughter of Jerusalem! See, your king comes to you righteous and having salvation, gentle and riding on a donkey, on a colt, the foal of a donkey."

Zechariah also provides this prophetic statement in Zechariah 12:10 (NIV): "And I will pour out on the house of

David and the inhabitants of Jerusalem a spirit of grace and supplication. They will look on me, the one they have pierced, and they will mourn for him as one mourns for an only child and grieve bitterly for him as one grieves for a firstborn son."

The Old Testament is filled with prophecies regarding the coming Messiah, yet when He arrived, the nation of Israel did not recognize Him. Matthew 16:1-3 (NIV) says, "The Pharisees and Sadducees came to Jesus and tested him by asking him to show them a sign from heaven. He replied, 'When evening comes, you say, 'it will be fair weather, for the sky is red; and in the morning, 'Today it will be stormy, for the sky is red and overcast.' You know how to interpret the appearance of the sky, but you cannot interpret the signs of the times." Jesus was referring to the signs and the prophecies that foretold His coming, the time they were then witnessing.

All the above prophecies regarding the life of the Jewish Messiah were fulfilled during the three-year period of Jesus' ministry to the nation of Israel. If the nation had recognized these events as they were occurring and had recalled the words of their prophets, they surely would not have missed His first coming. My prayer is that by God's grace, the Jewish people, God's chosen people, won't miss His second coming!

- 16 -

The Names of Adam's Descendants

Proverbs 25:2 (NIV) says, "It is the glory of God to conceal a matter; to search out a matter is the glory of kings." As seen up to this point, this book is filled with hidden messages that God has concealed in the Holy Scriptures and in His creation. The concealment of hidden messages is one of the things that makes the Bible unlike any other book ever written. It contains hundreds of prophecies, the majority of which have been fulfilled exactly as forecasted. However, there are many more that are yet to be fulfilled. This book has been written with an encrypted code to prevent the counterfeiting of the original text, and it unveils messages in three distinct formats: written words, pictograms that convey their own unique meaning, and gematria values that validate and authenticate the written narrative.

Another method that God used to convey a hidden message is found in the meanings behind the names of Adam's descendants. This message can be found in more detail in the YouTube video by the late Dr. Chuck Missler. Let's examine the amazing hidden message in the names of Adam's descendants, as provided in chapter 5 of the Book of Genesis. The message is revealed in the meanings of the names in chronological order, starting with Adam.

Name	Meaning
Adam	Man (is)
Seth	Appointed
Enosh	Mortal
Kenan	Sorrow, (but)
Mahalalel	The Blessed God
Jared	Shall come down
Enoch	Teaching
Methuselah	His death shall bring
Lamech	The Despairing
Noah	Comfort, (and) Rest

What we have is the gospel message of Jesus Christ embedded in the names of the genealogy from Adam to Noah: "Man is appointed mortal sorrow, but the Blessed God shall come down, teaching that his death shall bring the despairing comfort and rest." [59] As Dr. Chuck Missler states in his video, "If these words were in any other order, it wouldn't make sense. And here it is, a summary of the Christian gospel tucked away, here in the Torah in Genesis chapter 5. This has a lot of implications. First of all, there is no way you will ever convince me that a group of Jewish Rabbis contrived to hide a summary of the Christian gospel in a genealogy within their highly venerated Torah, the books of Moses. No way!" He went on to say, "But there is another implication here too, and that is that God's plan of redemption was not a knee-jerk reaction to the surprise that Adam blew it." He concludes his comments with a question, "When did God start dealing with you? Ephesians

1:4 tells you that before the foundation of the earth, God had you on His mind."

- 17 -

The Shroud of Turin

The Shroud of Turin is the most sacred and studied religious artifact in Christianity. [60] Many consider it to be the actual burial cloth of Jesus Christ. The cloth portrays a faint image of a man who was crucified. How the image got onto the cloth has been one of modern science's most perplexing and long-standing mysteries.

The Bible says that when Peter and John reached the tomb on the day of Jesus' resurrection, Peter went in and found linen cloths as well as a folded napkin that had been wrapped around His head. The Shroud of Turin is believed to be one of the linen cloths. The folded cloth, the Sudarium of Oviedo, is believed to be the napkin. Both the Shroud and napkin are thought to have been meticulously preserved for the past 2,000 years.

The bloodstains on the napkin are type AB and match those on the Shroud. The bloodstains on the head of the Shroud also align with those on the napkin. However, the most striking aspect of the Shroud is that the bloodstains replicate the biblical account of the scourging and crucifixion of Christ. There are bloodstains on the cloth where the hands and feet were nailed to the cross. There are additional bloodstains on the head where the crown of thorns was placed and a large bloodstain where His side was pierced. Lastly, there are bloodstains that cover the entire body, which resulted from the flogging that He endured before His crucifixion.

In 1948, a multi-disciplined team of scientists was given access to the Shroud to examine it for authenticity. The undertaking was known as the Shroud of Turin Research Project, and the team's task was to determine how the image got on the Shroud. The team took X-rays as well as ultraviolet and visual photos of the cloth. After detailed analysis, they were unable to determine how the image was made. For many, their research appeared to confirm the authenticity of the cloth since there was no explanation for how the image was formed. The scientific analysis of the image concluded that it was not created from paints, pigments, dyes, stains, acids, or powders.

The image on the cloth shows a body covered with scourge marks, like those that would be present after a beating or flogging. The nail marks on the cloth are on the wrists and feet rather than the hands, consistent with Roman crucifixions of that era. There is also a large bloodstain that appears to be from a wound on the man's side. This is consistent with the biblical account of the soldier who pierced Jesus' side to confirm he was dead.

Pollen grains and spores from plants native to Israel, specifically Jerusalem, were also found on the Shroud. [61] Furthermore, images of coins were found covering the eyes of the crucified man and provided additional confirmation regarding the timing of the crucifixion. The coin covering the right eye was a Lepton coin minted between 29 AD and 36 AD by Pontius Pilate. [62]

Another multi-disciplined team of experts examined the Shroud in 1978. This effort was called the Shroud of Turin Project. The team used microscopy, infrared spectrometry, X-ray fluorescence spectrometry, X-ray radiography, thermography, and ultraviolet fluorescence spectrometry and concluded that no material was added to the linen fibers to create the coloring observed in the image. [63] A photograph of the Shroud was processed using a VP-8 image analyzer. The photograph produced by the VP-8 image analyzer was even more puzzling, as the Shroud was found to contain holographic 3-dimensional information encoded within the image. [64]

In 1988, a piece of the Shroud was carbon-dated, and the selected material was determined to be from the 13th or 14th century AD. However, this test was later invalidated when it was determined that the edge pieces of the cloth used in the carbon dating test were not part of the original Shroud. A team of researchers sued the University of Oxford for the original data results and won. "After studying the data for two years, the new research team announced that the study from 1988 was flawed because it did not involve the study of the entire Shroud – just some edge pieces." [65] Research performed by Joe Mario and Sue Benford discovered that the edge pieces of the cloth used in the carbon dating analysis contained cotton fibers that were woven into the original linen. [66] "The repair was performed in 1532 after a fire in the chapel charred the cloth in several places." [67] This may account for the carbon dating of the cotton material as being from the 13th or 14th century.

Dr. Liberato De Caro used a more accurate dating method to determine the cloth's age and released the results of his peer-reviewed study in 2022. The methodology he used to date the Shroud was a wide-angle X-ray scattering (WAXS) technique, which determined its age to be at least 2,000 years. [68]

No other known historical artifacts display any of the unique characteristics found on the Shroud of Turin. [69] Until recently, scientists had no idea how the image was formed, nor had they been able to reproduce a similar image using today's technology. After multiple teams conducted detailed scientific analyses of the cloth, no one was able to provide a credible theory on how the image was produced until scientist Paola Di Lazzaro and her research team at Italy's ENEA (National Agency for New Technologies, Energy, and Sustainable Economic Development) conducted their experiments on the Shroud. She suggests that the image was made by an intense burst of ultraviolet light and was able to replicate how the image could have been produced. Using excimer lasers capable of producing extremely brief and intense pulses of UV light, Di Lazzaro's team successfully created superficial discolorations on linen that closely resembled the properties of the Shroud's image. [70] Di Lazzaro concluded that the amount of ultraviolet light needed to produce the image on the Shroud exceeds the maximum power released by all ultraviolet light sources available today. She determined that the burst of ultraviolet light required to form the image would be several billion watts and would have taken less than one forty-billionth of a second. A burst of ultraviolet light on linen cloth would turn single electron-bond carbon atoms into double electron-bond carbon atoms, changing their color and producing the image we see on the Shroud. [71]

A recap of the top reasons to suggest the Shroud of Turin is the actual burial cloth of Jesus Christ is provided below.

1. The Shroud of Turin is the world's most studied archaeological artifact, and until Paola Di Lazzaro suggested that a burst of ultraviolet light produced the image, no plausible explanation existed for how the image could have been created.

2. Some scientists now believe that a burst of radiation and ultraviolet light may have produced the image by transforming single electron-bond carbon atoms into double electron-bond carbon atoms.

3. The Shroud has been studied by scientists from multiple disciplines, and except for the faulty carbon-dating study, the age of the Shroud appears to correspond to the time of Christ.

4. The scientific investigative team concluded that there was no evidence of paint, pigment, dye, or any other material on the Shroud.

5. The radiocarbon dating in 1988 was unreliable.

6. The VP8 Imaging Analyzer depicts an encoded three-dimensional image.

7. Data from the Shroud indicates that it is from Israel at the time of Jesus' death.

8. John 20:7 says that the cloth was linen, and the Shroud is linen.

9. The blood on the cloth matches the historical account of Jesus' scourging and crucifixion.
    - There is blood where the hands and feet were pierced.
    - There is blood where the spear entered His side.
    - The largest blood stain is from the wound on the side.
    - The blood from the side is mixed with a clear, watery substance.
    - There is blood on His head where the crown of thorns would have been.
    - There are bloodstains from scourge marks all over His body.

10. The blood stains on the Shroud are consistent with those on the Sudarium of Oviedo.
    - This was the cloth that wrapped His head.
    - The blood on the Shroud and Sudarium of Oviedo matches; both are type AB.

11. There is a faint image of the Pilate coin over the right eye, dating the Shroud to 29-36 AD.

12. Dr. Liberato De Caro's 2022 peer-reviewed wide-angle X-ray scattering (WAXS) dating technique estimates the age of the Shroud to be approximately 2,000 years.

The following YouTube videos are recommended for additional information:

"How Image was formed on the Shroud of Turin FAST VERSION," by Good Shepard Films.

"NEW! Science Proves Shroud Image Is JESUS 2020 Video," by Good Shepard Films.

"Scientific Evidence on the Shroud of Turin," by the Joy of the Faith.

- 18 -

God Declares the End from the Beginning

The Hebrew language is unique in that it consists of three tiers of communication. The first uses letters of the alphabet to form words that convey meaning. The second tier is symbolism: each letter in the Hebrew alphabet has a pictogram that has its own distinct meaning. The third tier refers to the gematria values of the letters, words, sentences, or phrases, which often provide additional confirmation and authentication of the written word.

In Isaiah 46:9b-10 (NIV), there is a very interesting statement that says, "I am God, and there is no other; I am God, and there is none like me. *I make known the end from the beginning.*" In searching for clues to uncover where God has made known the end from the beginning, let's go to His Word, to the first word in the first sentence of the first chapter of the Bible. Is it possible that we could find the hidden message there, the message that makes known the end from the beginning? Is it possible that the pictograms and the numeric values of the letters in the first word contain a hidden message and the timeline for the end of this age?

The first Hebrew word in the Bible is Bereshit, which is spelled בראשית. This six-letter word means "In beginning" but is translated as "In the beginning" in our English Bible. Could it be that within the first six letters of the Bible, the Holy Spirit has embedded the message of man's redemption and salvation, as well as the timeline for the end of the age? Isaiah 46:9b-10 (NIV) says, " I am God, and there is no other; I am God, and there is none like me. *I make known the end from the beginning.*" Has God revealed the salvation message and three timelines within the first six letters of the Bible, in the word Bereshit?

The chart that follows shows the ancient symbols, or pictograms, for the 22 letters in the Hebrew alphabet. It also shows the numeric value of each letter and the meanings of the pictograms.

NUMERIC VALUES OF HEBREW LETTERs, PICTOGRAMS, AND MEANINGS					
1 - א Aleph	2 - ב Bet	3 - ג Gimel	4 - ד Dalet	5 - ה Hey	6 - ו Vav
Ox Head Power, Godhead	Tent, Floor-Plan Family, House, In, Into	Foot Gather, Walk, Carry	Tent Door Movement, Hang, Enter	Man Arms Raised Behold, Look, Reveal	Nail, Tent Peg Add, Secure, Hook
7 - ז Zayin	8 - ח Chet	9 - ט Tet	10 - י Yod	20 - כ Kaph	30 - ל Lamed
Mattock, Plough Weapon, Nourish	Tent Wall Outside, Devide, Half	Basket Surround, Contain	Hand and Arm Work, Throw, Worship, Deed	Open Hand Bend, Open, Allow, Tame	Shepard Staff Teach, Yoke, Toward, Bind
40 - מ Mem	50 - נ Nun	60 - ס Samech	70 - ע Ayin	80 - פ Pe	90 - צ Tsade
Water Chaos, Mighty, Blood	Sprouting Seed Continue, Heir, Son	Thorn Grab, Hate, Protect	Eye Watch, Know, Shade	Open Mouth Blow, Scatter, Edge	Destination Trail, Journey, Hunt
	100 - ק Qoph	200 - ר Resh	300 - ש Shin	400 - ת Tav	
	Sun at Horizon Condense, Circle, Time	Head, Authority, First, Leader	Teeth, Destroy Sharp, Press, Eat, Two, Fire	Cross, Mark Covenant, Sign, Signal	

Let's examine the Hebrew word Bereshit, בראשית, to determine if there is a hidden message in the pictograms, numeric values, and embedded words within these six letters. The first Hebrew letter in Bereshit is ב, Bet. The symbol for Bet is a house. The letter ב, Bet, is often used as a prefix to indicate the prepositions "in" or "into." Could it mean that someone has come out of their house (in heaven) and into the world?

The second letter is ר, Resh. The symbol for Resh is a head and means, authority, first, or leader. Could this letter be a reference to who came out of the house, a leader or head authority? The first two letters, Bet and Resh, also spell the Hebrew word "son," בר. Is it possible that the head authority who came out of the house is someone's son?

The third letter is א, Aleph. The symbol for Aleph is an ox head, which means strong, leader, power, or Godhead. It is often seen as a representation of God or Father. Looking at the first three letters in Bereshit, might we conclude that the son who came out of His house in heaven was God's son? These first three letters also form another Hebrew word, ברא, which means "created" or "Creator." Is it possible that God's Son is also the Creator? If so, it confirms what is said in John 1:3, "Through him all things were made; without him nothing was made that has been made."

The fourth letter in Bereshit is ש, Shin. The symbol for Shin is teeth, which means press, destroy, or fire. Amazingly, the second, third, and fourth letters in Bereshit also form another Hebrew word, ראש, which is the word

"Resh," meaning head, authority, first, or leader. Could this be a further confirmation of whom the prophecy is about and possibly shed light on the purpose of the son's coming?

The fifth letter in Bereshit is י, Yod. The symbol for Yod is an arm and hand, which can mean work or deed. This letter may suggest that the head authority, God's Son, will perform a deed by His own hand. Could that deed be voluntarily laying down His life for the sins of the world?

The sixth letter in Bereshit is ת, Tav. The symbol for Tav is a cross, and means mark, sign, or covenant. Could this pictogram of a cross represent the deed that would be performed? Could this sign also represent the covenant that the Son will make with those who choose to accept His deed, His sacrifice on the cross, and put their faith and trust in Him?

One way of interpreting the meaning of the symbols behind the six letters in the first word of the Bible would be: The Son of God will come out of His house in heaven to sacrifice His life, by His own hand, on a cross, in order to establish a new covenant with mankind. When considering the message from Isaiah 46:9b-10 (NIV), "I am God, and there is no other; I am God, and there is none like me. *I make known the end from the beginning,*" is it surprising to find that the first six Hebrew letters in the word of God are arranged in an order that may reveal God's plan of salvation and that this plan was established from the very beginning of time, from the time of the creation of the heavens and the earth?

Revelation 13:8 (KJV) says, "And all that dwell upon the earth shall worship him (Satan), whose names are not written in the book of life *of the Lamb slain from the foundation of the world.*" Again, we see that the plan of salvation was set in motion from the foundation of the world. Christ voluntarily sacrificed His own life for the sins of mankind from the foundation of the world, and the very first word in the Bible provides a pictographic illustration of the salvation message.

But there is more to this hidden message. There are also numeric values associated with these six letters. Is it possible that the numeric values of these letters also provide additional details regarding the timeline for when key events would occur and the timing for the end of the age? The first three chapters of Genesis contain the story of creation and the fall of man. If the first word in Genesis, Bereshit, בראשית, provides a pictographic account of the redemption plan of Christ, is it possible that the numeric values of the letters provide the timeline? Did God provide a prophetic 7,000-year timeline using the numeric values of the Hebrew letters in Bereshit?

In 2 Peter 3:8, it says, "But do not forget this one thing, dear friends: With the Lord a day is like a thousand years, and a thousand years are like a day." Do the numeric values of the six letters in Bereshit provide a timeline from Adam to Christ, from the cross to the beginning of His millennial reign, and from the cross to the end of His millennial reign? The crucifixion of Christ is believed to have taken place between 26 AD and 36 AD, during the reign of Pontius

Pilate, who governed Judea during that time. The dates most commonly held are 30 AD and 33 AD, because these dates seem to coincide fairly well with the events that occurred during the Passover week. However, I now believe the strongest evidence supports a crucifixion date of 32 AD, which would put His birth year at 2 BC. If this date is accurate, it would place the creation year at 4,002 BC, assuming the timeline is unfolding in 1,000-year periods.

The evidence supporting a crucifixion date of 32 AD follows. Daniel 9:24-26 (NKJV) says, "Seventy weeks are determined for your people and for your holy city, to finish the transgression, to make an end of sins, to make reconciliation for iniquity, to bring in everlasting righteousness, to seal up vision and prophecy, and to anoint the Most Holy. Know therefore and understand, that from the going forth of the command to restore and rebuild Jerusalem until Messiah the Prince, there shall be seven weeks and sixty-two weeks; the street shall be built again, and the wall, even in troublesome times. And after the sixty-two weeks Messiah shall be cut off, but not for Himself; and the people of the prince who is to come shall destroy the city and the sanctuary."

The command to restore and rebuild Jerusalem by King Artaxerxes Longimanus is believed to have occurred on 14 March 445 BC. [72] The seven weeks and sixty-two weeks refer to a period of 483 years, which equates to 173,880 days, using the 360-day Jewish calendar. This would be the number of days from Artaxerxes command until the date that Messiah would be cut off (crucified). It turns out to be

10 April or 14 Nisan in the year 32 AD on the Jewish calendar. Note that He is "cut off, but not for Himself." He was cut off, crucified, for the sins of the world. The day of preparation for the Passover was 10 April 32 AD, a Wednesday. The following day was the day of Passover, which fell on a Thursday. The next day was a Sabbath since every day that follows Passover is a Sabbath. There would have been two Sabbath days that week, Friday and Saturday.

What makes this date the most compelling, in my opinion, is that it is the only date that satisfies all the major prophecies regarding Jesus's death, burial, and resurrection. The date aligns perfectly with Daniel's prophecy of 483 years from the command to restore and rebuild Jerusalem until the death of the Messiah. Because Nisan 14 in the year 32 AD places the crucifixion date on a Wednesday, it satisfies Jesus' prophecy from Matthew 12:40 (NKJV), "For as Jonah was three days and three nights in the belly of the great fish, so will the Son of Man be three days and three nights in the heart of the earth." It keeps the resurrection day on Sunday and meets the criteria for the lunar cycle for Passover in the year 32 AD. No other crucifixion date satisfies all the criteria specified in the biblical prophecies.

Since the purpose of Christ's coming to the earth was to sacrifice Himself as the Passover Lamb, is it possible that the first 4,000 years represent the first four days on God's 7,000-year prophetic timeline? Is it also possible that the first 4,000 years on this prophetic timeline are counted from the day Adam and Eve sinned rather than from the day of creation? This would align the commission of the sin with the sacrifice for the sin 4,000 years later. If this is true, could the end of the church age then be counted from the time of His death on the cross rather than from the time of His birth? That would place the end of the church age somewhere around 2032, assuming His death occurred in 32 AD.

Romans 5:15 (NIV) says, "But the gift is not like the trespass. For if the many died by the trespass of the one man (Adam), how much more did God's grace and the gift that came by the grace of the one man, Jesus Christ, overflow to the many!" I Corinthians 15:22 (NIV) goes on to say, "For as in Adam all die, so in Christ all will be made alive." And finally, in I Corinthians 15:45 (NIV), we read, "So it is written: 'The first man Adam became a living being'; the last Adam (Jesus), a life-giving spirit." Is it possible that the first Adam lived a sinless life for the exact same time that Jesus lived a sinless life, 33 and a half years? If so, the time from Adam's first sin to Jesus' sacrificial death to pay for that sin would be exactly 4,000 years. This would align the time of the sin with the need for a sin-bearer. Since the penalty for sin is death, Christ's death on the cross paid for Adam's sin and for all those who put their trust and faith in Him.

Genesis 2:16-17 (NIV) says, "And the LORD God commanded the man, 'You are free to eat from any tree in the garden; but you must not eat from the tree of the knowledge of good and evil, for *when* you eat of it you will surely die.'" Now, let's examine the 7,000-year biblical timeline for man's time on earth and see if the key dates can be derived from the numeric values in the first word of the Bible. The values for the letters in Bereshit are shown below. The key to unlocking the mystery of when key events take place may be tied to the value of the letter Yod (10), which some have suggested is God's divine multiplier. [73]

Here are the first six Hebrew letters in the word Bereshit and their numeric values.

TAV	YOD	SHIN	ALEF	RESH	BEYT
ת	י	שׁ	א	ר	ב
400	10	300	1	200	2

If we examine the last two letters in the word Bereshit, the letters Yod י and Tav ת, we may find our first clue regarding the timing of the most important event in history: the crucifixion of Christ. The symbol for Yod י is a pictogram of a hand and arm, while the symbol for Tav ת is a pictogram of a cross. If we multiply the value of Tav (400) by the divine multiplier Yod (10), we get 4,000 years from the first Adam's sin to the second Adam's sacrificial death on the cross. This may be the most significant prophecy contained in the word Bereshit. The other two prophecies on this timeline relate to the church age and the end of the millennial reign of Christ. Both timelines can be derived from the date of the crucifixion, which may be 32 AD.

The Bible says that Jesus will return to establish His kingdom on earth and that He will reign as the "King of kings" and the "Lord of lords" for a thousand years. We may be able to determine the timing of this prophecy by multiplying the value of Resh (200), which represents head, authority, first, or leader, by the divine multiplier Yod (10). The result is 2,000 years. Could this represent the church age, and suggest that the millennial reign of Christ will begin 2,000 years after His death on the cross?

The final Bereshit prophecy foretells the destruction of the world by fire. The third and fourth letters in the word Bereshit are אש, which is the Hebrew word for "fire." The letter Shin ש can mean destroy or fire. When we multiply the value of Shin ש (300) by the divine multiplier, Yod (10), we may have the prophetic timeline for when the destruction of the world by fire will occur. The product of this multiplication, 300 x 10, equals 3,000, which may indicate the number of years from Jesus' death on the cross to the end of the millennial reign of Christ.

If the above assumptions are true, the timeline might resemble the following chart.

Seven Thousand Year Timeline Derived from the Bereshit Prophecy

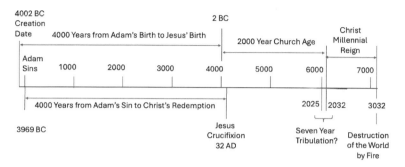

Revelation 20:7 (NIV) says, "When the thousand years are over, Satan will be released from his prison and will go out to deceive the nations in the four corners of the earth – Gog and Magog – to gather them for battle." 2 Peter 3:7 says, "By the same word the present heavens and earth are reserved for fire, being kept for the day of judgment and destruction of ungodly men." 2 Peter 3:12b (NIV) goes on to say, "That day will bring about the destruction of the heavens by fire, and the elements will melt in the heat." After this, God promises to create a new heaven and a new earth. Revelation 21:1 (NIV) says, "Then I saw a new heaven and a new earth, for the first heaven and the first earth had passed away..." This new heaven and earth will be reserved for the righteous saints of God.

For additional details on the Bereshit timeline and an alternative crucifixion date, reference any of the YouTube videos by RockislandBooks, such as, "Is the End of Days Prophesied in the First Word of the Bible?"

- 19 -

Conception and the Spark of Life

A team of scientists at Northwestern University discovered that a flash of light occurs when a mammal's sperm cell penetrates an egg. The event was first observed in mice in 2011. In 2014, they filmed this flash of light for the first time. [74] What the scientists witnessed was billions of zinc atoms being released at the exact moment the sperm penetrated the egg. The team found that each egg contained about 8,000 zinc compartments with about 1 million zinc atoms in each compartment. At the moment of conception, billions of zinc atoms are released, producing a spark of light and shutting down all entry ports for sperm to access the egg. The team making this discovery and providing the above information was headed by Teresa Woodruff, Director of the Women's Health Research Institute and Chief of the Division of Obstetrics and Gynecology-Fertility Preservation at Northwestern University. [75] To see the flash of light that occurs at the moment of conception, view the YouTube video titled: "Watch the moment the egg is fertilized" by Hashem Al-Ghaili.

In John 8:12 (NIV), Jesus said, "I am the light of the world. Whoever follows me will never walk in darkness but will have the light of life." It's interesting that Jesus infers a correlation between light and life. Before 2011, science did not know that there was a burst of light when conception occurred. John 1:4 (NIV) says, "In him was life, and that life was the light of men." Is it possible that there is a connection between the spark of light that occurs at conception and life itself? In John 6:33, we read, "For the bread of God is the bread that comes down from heaven and gives life to the world." And in John 8:12 (NIV), it says, "I am the light of the world. Whoever follows me will never walk in darkness but will have the light of life." Is it possible that the light of life that Jesus spoke of is a reference to the spark of light that occurs at the moment of conception, the moment life comes into existence? Is this another sign of the reliability of His Word?

The Bible describes Jesus as the "light of the world." Colossians 1:16 (NIV) says, "For in him all things were created: things in heaven and on earth, visible and invisible, whether thrones or powers or rulers or authorities; all things have been created through him and for him." Is it surprising that Jesus, the "light of the world," would begin each new life—whether human or animal—with an explosion of light? Could this also be the moment that the spirit enters the new creation?

Is there a connection between the spark of life that occurs at conception and the departure of the spirit from the body at the moment of death? It is still a mystery what transpires when life leaves the body since all the material components of the cells that make up the body are still present at the time of death. What exactly changed materially at the moment of death? Could it be that rather than a material change, something happened spiritually? Is it possible that the spiritual being, whose residence was the body, suddenly departed? In the movie "After Death" and in the book "Life After Life," there are numerous case studies of individuals who were clinically dead but later revived. In some cases, the person had been dead for up to half an hour; both the brain and heart had stopped functioning, and yet, somehow, they came back. Even more astonishing, when some of them returned, they provided detailed descriptions of what took place in the operating room while they were clinically dead. Many of these individuals also provided detailed accounts of heaven or hell, which they say they visited during the time they were pronounced dead. For more information on after-death experiences, the book *Life After Life*, by Raymond Moody, provides further details on documented near-death experiences.

– 20 –

Bible Truths That Predate Science

This chapter explores Scriptures containing scientific facts that predate their scientific discovery. The fact that the Bible provided this information before the scientific community discovered it should cause one to ask, "Who or what was the source of the information provided in the Bible?" From where did the Bible get these facts before the rest of the world discovered them?

When reading the opening verses in the Bible, Genesis 1:1-3, most people do so without giving them much thought. However, a close examination of these verses reveals that the author of Genesis used the same fundamental principles that today's scientists use to describe the composition of our universe. In describing the universe, modern science employs five very specific terms: time, space, matter, energy, and motion. Genesis 1:1-3 (KJV) describes the creation of the universe with virtually the same words: "In the beginning (time) God created (energy) the heavens (space) and the earth (matter)... and the Spirit of God moved (motion) upon the face of the waters." This description of our universe did not exist before the last two centuries. It took nearly six thousand years for science to catch up.

The phrase used in Genesis 1:1, "In the beginning," refers specifically to the beginning of time. However, for thousands of years, scientists were convinced that the universe had always existed and was eternal. Then along came Georges Lemaître, a Belgian cosmologist, who proposed the Big Bang Theory in a paper he wrote in 1931. [76] The Big Bang Theory was one of the first to suggest that the universe had a beginning. This theory was widely acclaimed and is still one of the most widely held beliefs regarding the origin of the universe. However, the Bible made this same claim thousands of years before science reached the same conclusion. A December 27, 1976, Time Magazine article titled "STARS Where Life Begins" said, "Most cosmologists – scientists who study the structure and evolution of the universe – agree that the biblical account of creation, in imagining an initial void, may be uncannily close to the truth." [77]

Consider this verse from Isaiah 42:5 (KJV), "Thus saith God the LORD, he that created the heavens, and stretched them out..." It wasn't until 1925 that Edwin Hubble, an American astronomer, proved that the universe is expanding and that there's a relationship between the speed at which distant galaxies are moving and their distance from the Earth. Those that are furthest away are moving the fastest. [78] We know that they are moving away from us because the light from distant galaxies is shifted toward the red spectrum. But once again, the author of the Bible knew this when He stated, "...the Creator of the heavens...stretches them out..."

Here's another verse from the Bible that predates the scientific discovery: Job 26:7 (NIV) states, "He spreads out the northern skies over empty space..." Astronomers have discovered a massive hole (empty space) in our universe. It's nearly a billion light-years in diameter and is located in the Northern Hemisphere near the constellation Boötes. [79] How did the author of the Bible know that there was empty space in the northern skies before there were telescopes?

Here's another example from the Book of Job. For most of man's history, it was believed that the Earth was flat, and many cultures believed that the Earth was supported on the back of a turtle. But Job 26:7b (NIV), which some scholars believe is the oldest book in the Bible, proclaimed, "He (God) suspends the Earth over nothing." At the time Job wrote these words, there was no knowledge of gravity, our solar system, or the rotation and orbits of the planets. So, where did this knowledge come from?

Another example from Job 28:25 (KJV) says, "To make the weight for the winds: and He weigheth the waters by measure." Some references suggest that Aristotle may have been the first to claim that air has weight, but most scholars believe that the first proof wasn't provided until the mid-1600s. Some give credit for this discovery to Evangelista Torricelli and Essais Rey. Others suggest Galileo. [80] But well before any of these men had any idea that air had weight, the Bible recorded this fact.

We read in Jeremiah 33:22 (KJV), "As the host of heaven cannot be counted..." This is a reference to the fact that the stars in heaven are too numerous to count. However, this would not have been the perspective of anyone living before the invention of Galileo's telescope in 1609. Only with the later invention of the Hubble telescope did we learn that the host of heaven (the stars) was far beyond our ability to count. Astronomy.com, in a September 28, 2021, article, estimated that there are 2 trillion galaxies, and they estimated that the number of stars is 200 sextillion (200,000,000,000,000,000,000,000). [81]

Here's a Bible fact that was only recently discovered. Hebrews 11:3 (NIV) says, "By faith we understand that the universe was formed at God's command, so that what is seen was not made out of what was visible." In recent years, science has made amazing breakthroughs in the field of subatomic particle research and quantum physics and now believes that nothing is solid. Everything in the universe is made up of energy. [82] For most of human history, mankind had virtually no knowledge of the realm of subatomic particles since they were well beyond our ability to see. Where did this information and knowledge about the world come from, suggesting that everything we see is made up of things that are not visible (energy)? How did the author of the Book of Hebrews know this before science did?

We read in Psalm 19:4-6 (KJV), "Their line has gone out through all the earth, and their utterances to the end of the world. In them He has placed a tent for the sun, which is as a bridegroom coming out of his chamber; it rejoices as a strong man to run his course. Its rising is from one end of the heavens, and its *circuit* to the other end of them, and there is nothing hidden from its heat." Before 1543, scientists believed that the Earth was the center of our universe and that the Sun revolved around the Earth. Nicolaus Copernicus, the father of modern astronomy, was the first to postulate that the Earth revolved around a stationary Sun. It was not until the early 1900s that Harlow Shapley discovered and measured the Sun's orbit around the Milky Way galaxy, proving the ancient Scriptures to be, once again, far ahead of science in declaring the Sun's circuit through the heavens. We now understand that our Sun and our entire solar system orbit the Milky Way galaxy every 230 million years. [83] That's quite a circuit, from one end of the heavens to the other end.

Here's an example of the Bible predating science on matters of health. Numbers 19:11-12 (NIV) says, "Whoever touches the dead body of anyone will be unclean for seven days. He must purify himself with the water on the third day and on the seventh day; then he will be clean." The waters of purification are defined in Numbers 19:6 (NIV), which says, "The priest is to take some cedar wood, hyssop, and scarlet wool and throw them onto the burning heifer." Modern science now knows that this mixture of items works very well in purifying substances or killing germs. Heifer ashes and cedar wood can be used to make lye. The hyssop plant converts into thymol (isopropyl alcohol), which kills bacteria. [84] Scarlet wool, similar to a Brillo pad, forms a gritty substance for scrubbing and cleaning the surface of the skin. Applying this substance on the third and seventh days allows it time to dry. Bacteria grow in a damp environment, so reapplying this after three and seven days helps ensure that all the bacteria are killed. Where did Moses get this medical knowledge thousands of years before modern science discovered bacteria and germs?

Another example of the Bible predating modern medical science is found in Genesis 17:12 (NIV), which reads, "For the generations to come every male among you who is eight days old must be circumcised..." Why did Moses say it should be the eighth day? There are two elements in our blood that are necessary for clotting: one is vitamin K, and the other is prothrombin. Vitamin K develops in a newborn between days 5 and 7. Prothrombin peaks on day eight and is never as high for the rest of your life. [85] So, day eight looks like the best day for a surgical procedure on a newborn to ensure clotting occurs. How did Moses know this before medical science understood it? Today, males circumcised shortly after birth are given vitamin K to assist with clotting.

This next example is interesting. Leviticus 11:6 (NIV) says, "The rabbit, though it chews the cud, does not have a divided hoof; it is unclean for you." This Scripture has been cited as an example of an error in the Bible since rabbits don't chew the cud, at least not in the same manner as other animals, like cows, goats, and antelope. This Scripture reference appears to be inaccurate on the surface. However, when we examine the original Hebrew text, we find that the word translated as "cud" is the Hebrew word *gêrâh*, which means something that has been swallowed. [86] Using the original inspired text, we find the statement to be true. The rabbit produces two different types of pellets after consuming food. The first is a light brown or greenish pellet that is edible. The second is darker in color and is the final waste product. The light brown or greenish pellet, which was previously swallowed, is eaten again. The Bible once again proves to be scientifically accurate regarding rabbits consuming something previously swallowed.

Here's another example of Bible knowledge predating scientific discovery by thousands of years. Genesis 6:5 (KJV) says, "And God saw that the wickedness of man was great in the earth, and that every imagination of *the thoughts of his heart* was only evil continually." And Job 17:11 (KJV) reads, "My days are past, my purposes are broken off, even the *thoughts of my heart.*" And Matthew 15:19 (KJV) states, "For *out of the heart proceed evil thoughts*, murders, adulteries, fornications, thefts, false witness, blasphemies." For more than four thousand years, almost everyone who has read the above Scriptures regarding *the thoughts of the heart* likely assumed these words were a metaphor.

The above Scriptures are an example of how the word of God can be taken literally. If the Holy Spirit inspired holy men and prophets to pen the words *"thoughts of the heart,"* could it be that the heart can have thoughts? That's exactly what Dr. J. Andrew Armour discovered. He introduced the term "heart brain" in 1991 and proclaimed that the heart has a complex and intrinsic nervous system—that is, a brain. [87] What follows is a list from the website heartmath.org of some of the attributes or characteristics of the little brain within the heart.

- The heart sends emotional and intuitive signals to help govern our lives.
- The heart directs and aligns systems in the body.
- The heart communicates with the brain, constantly.
- The heart makes many of its own decisions.
- The heart has an independent complex nervous system known as "the brain in the heart." [88]

According to an April 24, 2011, article in "NAMAH, the Journal of New Approaches to Medicine and Health," there are several stories of individuals experiencing personality changes after receiving a heart transplant. [89] According to the article, some heart transplant recipients had memories of their donors. The article begins by saying, "There have been perplexing reports of organ transplant receivers claiming that they seem to have inherited the memory, experiences, and emotions of their deceased donors, which are causing quirky changes in their personality." The article gives an account of an eight-year-old girl who was having nightmares about the murder of her ten-year-old donor. The girl's parents took her to the police, where she described the details of the murder. The information was used by the police to identify and arrest the perpetrator. [90]

- 21 -

God's Prophets Foretell the Future

One of the more unique characteristics of the Bible is its prophetic content: its ability to predict the future. The Bible has prophesied hundreds of events decades or even centuries before they occurred. This could only be possible if the author was God. 2 Timothy 3:16 (KJV) asserts that the author was God when it says, "All Scripture is given by inspiration of God…"

The previous chapters in this book have shown how God, the Holy Spirit, has encoded all Scripture with a numeric seal that could not have been constructed by man since the numeric values of the Hebrew alphabet were not established until long after the Old Testament was written. According to Wikipedia, the numeral system using the letters of the Hebrew alphabet was adapted from the Greek numerals sometime between 200 and 78 BCE. [91]

Even though the numeric values of the Greek alphabet were in place before the New Testament was written, the complexity of the numeric patterns found in all 27 books is believed to be beyond human capability. Ivan Panin made the following statement regarding the astronomical improbabilities of the patterns of 7, 37, and other numbers that he discovered in the Holy Scriptures: "…the Bible could not possibly have been written by mere human beings alone, but that it is supernatural, God inspired, God-given book." [92] 2 Timothy 3:16 (KJV) confirms Dr. Ivan Panin's conclusion that "All Scripture is given by inspiration of God."

One of the strongest evidences supporting the divine origin of the Holy Scriptures is the accuracy of the prophetic words spoken by God's prophets. Not only did their words predict events in the Old and New Testaments before they happened, they also foretold events that are yet to occur. According to "Christianity: Logic and Science," there are 8,352 prophetic passages in the Bible, making 1,817 predictions of future events that cover over 700 topics. [93] Other than God, how else would we explain how the prophets in the Old and New Testaments were able to predict future events and specific details regarding those events hundreds or thousands of years before they occurred? Where else could this information have come from, if not from an omniscient, omnipotent, and omnipresent God—the one who dwells outside of time, space, and matter and sees the end from the beginning?

It is well established that the Old Testament is filled with Messianic prophecies regarding the life and death of Jesus. God's prophets, inspired by the Holy Spirit, not only foretold the coming of the Messiah but also provided minute details regarding His birthplace, family heritage, where He would live, what He would do, why He came, and how He would die. What follows is an abbreviated list of Old Testament prophecies and their fulfillment, as described in the New Testament. Keep in mind that the original prophecies were given hundreds of years before the actual events occurred. Jesus could not possibly have controlled the outcome of most of these events, even if He wanted to.

In the following prophecies, OT refers to Old Testament prophecies, and NT refers to their New Testament fulfillment.

Prophecy: He would be born in Bethlehem of the tribe of Judah and his origins would be from ancient times.

OT: Micah 5:2 (KJV), "But thou, Bethlehem Ephrathah, though you are small among the tribes of Judah, out of you will come for me one who will be ruler over Israel, whose origins are from of old, from ancient times."

NT: Matthew 2:1 (NIV), "...Jesus was born in Bethlehem in Judea..."

NT: John 8:58 (NIV) "I tell you the truth," Jesus answered, "before Abraham was born, I am!"

Yet to be fulfilled in the above prophecy is His rule over Israel. In 1 Tim 6:13-15 (NIV), we read, "In the sight of God, who gives life to everything, and of Christ Jesus, who while testifying before Pontius Pilate made the good confession, I charge you to keep this commandment without spot or blame until the appearing of our Lord Jesus Christ, which God will bring about in His own time—God, the blessed and only Ruler, the King of kings and Lord of lords..." He will rule over Israel and the world.

Prophecy: He would be born of a virgin.

OT: Isaiah 7:14 (KJV), "Therefore the Lord himself shall give you a sign; Behold, a virgin shall conceive, and bear a son, and shall call his name Immanuel." [Meaning, God with us]

NT: Matthew 1:18 (NIV), "This is how the birth of Jesus Christ came about. His mother Mary was pledged to be married to Joseph, but before they came together, she was found to be with child through the Holy Spirit."

Prophecy: He would perform healing miracles.

OT: Isaiah 35:4-6 (KJV), "...he will come to save you. Then will the eyes of the blind be opened and the ears of the deaf unstopped. Then will the lame leap like a deer, and the tongue of the dumb shout for joy."

NT: Matthew 11:4-5 (NIV), "Jesus replied, 'Go back and report to John what you hear and see: The blind receive sight, the lame walk, those who have leprosy are cured, the deaf hear, the dead are raised, and the good news is preached to the poor."

Prophecy: He would be betrayed by a friend.

OT: Psalm 41:9 (NIV), "Even my close friend, whom I trusted, he who shared my bread has lifted up his heel against me."

NT: Matthew 26:23 (NIV), "Jesus replied, 'The one who has dipped his hand into the bowl with me will betray me.'"

Prophecy: The Messiah would be betrayed for 30 pieces of silver.

OT: Zechariah 11:12 (KJV), "And I said to them, 'If it is agreeable to you, give me my wages; and if not, refrain.' So they weighed out for my wages thirty pieces of silver."

NT: Matthew 26:14-15 (KJV), "Then one of the twelve, called Judas Iscariot, went to the chief priests, and said, 'What are you willing to give me if I deliver Him up to you?' And they counted out to him thirty pieces of silver."

Prophecy: The money received for his betrayal would be cast to the potter.

OT: Zechariah 11:13 (NIV), "And the LORD said to me, 'Throw it to the potter' – the handsome price at which they priced me! So I took the thirty shekels of silver and threw them into the house of the Lord to the potter."

NT: Matthew 27:4-7 (NIV) provides the account of Judas attempting to return the silver to the chief priests, "'I have sinned,' he said, 'for I have betrayed innocent blood.' 'What is that to us they replied. That's your responsibility.' So

Judas threw the money into the temple and left. Then he went away and hanged himself. The chief priests picked up the coins and said, 'It is against the law to put this into the treasury, since it is blood money.' So they decided to use the money to buy the potter's field as a burial place for foreigners.'"

Prophecy: He would be forsaken by his disciples.

OT: Zechariah 13:7b (NIV), "Strike the Shepherd that the sheep may be scattered…"

NT: Matthew 26:56 (KJV), "But all this was done, that the Scriptures of the prophets might be fulfilled. Then all the disciples forsook him and fled."

Prophecy: He would be accused by false witnesses.

OT: Psalm 35:11 (KJV), "False witnesses did rise up; they laid to my charge things that I knew not."

NT: Matthew 26:59-60 (KJV), "Now the chief priests, and elders, and all the council, sought false witness against Jesus, to put him to death; But found none. At the last came two false witnesses."

Prophecy: He would be smitten and spat upon.

OT: Isaiah 50:6 (NIV), "I offered my back to those who beat me, my cheeks to those who pulled out my beard; I did not hide my face from mocking and spitting."

NT: Matthew 27:30 (KJV), "And they spit upon him, and took the reed, and smote him on the head."

Prophecy: He would be silent before his accusers.

OT: Isaiah 53:7 (NIV), "He was oppressed and afflicted, yet he did not open his mouth; he was led like a lamb to the slaughter, and as a sheep before her shearers is silent, so he did not open his mouth."

NT: Matthew 27:12-14 (NIV), "When he was accused by the chief priests and the elders, he gave no answer. Then Pilate asked him, 'Don't you hear how many things they are accusing you of?' But Jesus made no reply, not even to a single charge–to the great amazement of the governor."

Prophecy: He would be wounded and pierced for our transgressions.

OT: Isaiah 53:5 (NIV), "But he was pierced for our transgressions, he was crushed for our iniquities; the punishment that brought us peace was upon him, and by his wounds we are healed."

NT: 1 Peter 2:24 (NIV), "He himself bore our sins in his body on the tree, so that we might die to sins and live for righteousness; by his wounds you have been healed."

NT: John 19:34 (KJV), "But one of the soldiers with a spear pierced his side, and forthwith came there out blood and water."

Prophecy: His hands and feet would be pierced.

OT: Psalm 22:16 (NIV), "For dogs have surrounded me; A band of evildoers has encompassed me; They pierced my hands and my feet." Note: David, the writer of Psalm 22, wrote these words approximately 500 years before the earliest recorded history of crucifixions (Herodotus 522 BC). [94]

NT: Luke 23:33 (NIV), "When they came to the place called The Skull, there they crucified him, along with the criminals – one on the right, the other on the left."

Prophecy: The people would wag their heads at him.

OT: Psalm 109:25 (RSV), "I am an object of scorn to my accusers; when they see me, they wag their heads."

NT: Matthew 27:39 (KJV), "And they that passed by reviled him, wagging their heads..."

Prophecy: The people would ridicule him, saying if he is the Son of God, let him save himself.

OT: Psalm 22:7-8 (NIV), "All who see me mock me; they hurl insults, shaking their heads: 'He trusts in the LORD; let the LORD rescue him. Let him deliver him, since he delights in him.'"

NT: Matthew 27:41-43 (NIV), "In the same way the chief priests, the teachers of the law and the elders mocked him.

'He saved others,' they said, 'but he can't save himself! He's the king of Israel! Let him come down now from the cross, and we will believe in him. He trusts in God. Let God rescue him now if he wants him, for he said, 'I am the Son of God.'"

Prophecy: They would divide his garments and cast lots for his clothing.

OT: Psalm 22:18 (NIV), "They divide my garments among them, and cast lots for my clothing."

NT: John 19:23-24 (NIV), "When the soldiers crucified Jesus, they took his clothes, dividing them into four shares, one for each of them, with the undergarment remaining. This garment was seamless, woven in one piece from top to bottom. 'Let's not tear it,' they said to one another. 'Let's decide by lot who will get it.'"

Prophecy: He would be forsaken by God

OT: Psalm 22:1 (NIV), "My God, my God, why have you forsaken me?"

NT: Matthew 27:46 (NIV), "About the ninth hour Jesus cried out in a loud voice, 'Eli, Eli, lama sabachthani?' – which means, 'My God, my God, why have you forsaken me?'"

Prophecy: He would be given gall and vinegar.

OT: Psalm 69:21 (NIV), "They put gall in my food, and gave me vinegar for my thirst."

NT: Matthew 27:34 (KJV), They gave him vinegar to drink mingled with gall: and when he had tasted thereof, he would not drink."

Prophecy: He would commit his spirit to God.

OT: Psalm 31:5 (KJV), "Into Your hand I commit my spirit; You have redeemed me, O LORD God of truth."

NT: Luke 23:46 (NIV), "Jesus called out with a loud voice, 'Father, into your hands I commit my spirit.' When he had said this, he breathed his last."

Prophecy: His loved ones and friends would stand afar off.

OT: Psalm 38:11 (NIV), "My friends and companions avoid me because of my wounds; my neighbors stay far away."

NT: Luke 23:49 (NIV), "But all those who knew him, including the women who had followed him from Galilee, stood at a distance, watching these things."

Prophecy: Not one of His bones would be broken.

As Jesus was the Passover Lamb, sacrificed for the sins of the world, it was important that He fulfill the Passover restrictions provided by Moses in Exodus 12:46b (NIV), "Do not break any of the bones." This was one of many prophecies that neither he nor his disciples could have controlled.

OT: Psalm 34:20 (NIV), "...he protects all his bones, not one of them will be broken."

NT: John 19:33 (NIV), "But when they came to Jesus and found that he was already dead, they did not break his legs." [As was the Roman custom for crucifixions.]

Prophecy: His side would be pierced.

OT: Zechariah 12:10 (NIV), "And I will pour out on the house of David and on the inhabitants of Jerusalem, the Spirit of grace and of supplication. They will look on me whom they have pierced, and they will mourn for him as one mourns for an only child, and grieve bitterly for him as one grieves for a firstborn son."

NT: John 19:34 (NIV), "...one of the soldiers pierced Jesus' side with a spear, bringing a sudden flow of blood and water."

Prophecy: Darkness would come over the land at the sixth hour.

OT: Amos 8:9 (NKJV), "'And it shall come about in that day,' says the Lord GOD, 'That I shall make the sun go down at noon, And I will darken the earth in broad daylight;"
NT: Matthew 27:45 (NIV), "From the sixth hour until the ninth hour darkness came over all the land." [The sixth

hour was around noon and the ninth hour was about three o'clock.]

Prophecy: He would be buried in a rich man's tomb.

OT: Isaiah 53:9 (NIV), "He was assigned a grave with the wicked, and with the rich in his death, though he had done no violence, nor was any deceit in his mouth."

NT: Matthew 27:57-60 (NKJV), "Now when evening had come, there came a rich man from Arimathea, named Joseph, who himself had also become a disciple of Jesus. This man went to Pilate and asked for the body of Jesus. Then Pilate commanded the body to be given to him. When Joseph had taken the body, he wrapped it in a clean linen cloth, and laid it in his new tomb which he had hewn out of the rock; and he rolled a large stone against the door of the tomb, and departed." [Had Joseph of Arimathea not asked for his body and laid him in his own tomb, Jesus would have been assigned a grave with the other two crucified criminals]

All the above prophecies came to pass exactly as prophesied. It has been conservatively estimated that Jesus fulfilled around 300 Old Testament prophecies during His ministry. [95]

Only God, who sees the end from the beginning and lives outside of time, space, and matter, could have provided such precise details of the events that would take place hundreds of years in the future. Only God could have communicated to His prophets the specific actions and the words that would be spoken by the men who cast lots for Jesus' clothing. We read in 2 Peter 1:20-21 (NIV), "Above all, you must understand that no prophecy of Scripture came about by the prophet's own interpretation. For prophecy never had its origin in the will of man, but men spoke from God as they were carried along by the Holy Spirit."

Imagine if you were told that a person who was born in the 1800s predicted that a man named Ronald Reagan would become a movie star, Governor of the state of California, and later President of the United States. Most would conclude that this prophecy was a myth and that the story was concocted by a scam artist sometime after Reagan became President. However, history clearly shows that just such an event did occur about 200 years after the prophet Isaiah foretold that a future king named Cyrus would conquer Babylon, restore the Jewish people to their homeland, and allow them to rebuild their temple.

Most scholars believe that Isaiah lived between 740 BC and 690 BC, and that King Cyrus was born between 590 BC and 580 BC. [96] The following Scriptures from the Book of Isaiah were written more than 100 years before King Cyrus was born and were fulfilled during his reign as king of the Persian-Median Empire.

The Cyrus Prophecies

OT: Isaiah 44:28 (NIV), "...who says of Cyrus, 'He is my shepherd and will accomplish all that I please;' he will say of Jerusalem, 'Let it be rebuilt,' and of the temple, 'Let its foundations be laid.'"

OT: Isaiah 45:1 (NIV), "This is what the Lord says to his anointed, to Cyrus, whose right hand I take hold of to subdue nations before him and to strip kings of their armor, to open doors before him so that gates will not be shut:"

The following Scriptures regarding the fulfillment of Isaiah's prophecies are from the books of 2 Chronicles and Ezra. The Book of 2 Chronicles was written around 430 BC. [97] The author is believed to be Ezra. The Book of Ezra, "Composed around 450 BC, though perhaps started earlier, ...documents the events that occurred between 538-450 BC." [98]

OT: 2 Chronicles 36:22-23 (NIV) states, "In the first year of Cyrus king of Persia, in order to fulfill the word of the LORD spoken by Jeremiah, the LORD moved the heart of Cyrus king of Persia to make a proclamation throughout his realm and to put it in writing. This is what Cyrus king of Persia says: "The LORD, the God of heaven, has given me all the kingdoms of the earth and he has appointed me to build a temple for him at Jerusalem in Judah. Any of his people among you may go up and may the LORD their God be with them."

OT: Ezra 1:1 (NIV), "In the first year of Cyrus king of Persia, in order to fulfill the word of the Lord spoken by Jeremiah, the Lord moved the heart of Cyrus king of Persia to make a proclamation throughout his realm and also to put it in writing:"

OT: Ezra 6:3 (NIV), "In the first year of King Cyrus, the king issued a decree concerning the temple of God in Jerusalem: Let the temple be rebuilt as a place to present sacrifices, and let its foundations be laid."

Now let's look at a different type of prophetic word given to King David by the prophet Nathan. It shows just how clearly God the Holy Spirit speaks to His servants and how He reveals to His prophets hidden things that only God could know. In the Book of 2 Samuel, chapter 11, we read the account of David and Bathsheba, where David has an affair with Bathsheba and conceives a son. When she tells David she's pregnant, he sends for her husband Uriah, who is fighting in a war with the Ammonites, and asks him how the war is going. David tells Uriah to go down to his house and wash his feet, but Uriah refuses. Uriah feels it wouldn't be right for him to be at home with his wife while the army of Israel is at war. David invites Uriah to eat and drink with him in an attempt to get Uriah drunk. But he still won't go home to his wife. Finally, he gives him a letter to deliver to Joab, the commander of David's forces. The letter instructs Joab to place Uriah at the front where the fighting is fiercest and then have the troops draw back so that Uriah is killed.

In 2 Samuel 12:1-18 (NIV), we read, "The Lord sent Nathan (the prophet) to David. When he came to him, he said, 'There were two men in a certain town, one rich and the other poor. The rich man had a very large number of sheep and cattle, but the poor man had nothing except one little ewe lamb he had bought. He raised it, and it grew up with him and his children. It shared his food, drank from his cup and even slept in his arms. It was like a daughter to him. Now a traveler came to the rich man, but the rich man refrained from taking one of his own sheep or cattle to prepare a meal for the traveler who had come to him. Instead, he took the ewe lamb that belonged to the poor man and prepared it for the one who had come to him.' David burned with anger against the man and said to Nathan, 'As surely as the Lord lives, the man who did this must die! He must pay for that lamb four times over, because he did such a thing and had no pity.' Then Nathan said to David, 'You are the man! This is what the Lord, the God of Israel, says:' 'I anointed you king over Israel, and I delivered you from the hand of Saul. I gave your master's house to you, and your master's wives into your arms. I gave you all Israel and Judah. And if all this had been too little, I would have given you even more. Why did you despise the word of the Lord by doing what is evil in his eyes? You struck down Uriah the Hittite with the sword and took his wife to be your own. You killed him with the sword of the Ammonites. Now, therefore, the sword will never depart from your house, because you despised me and took the wife of Uriah the Hittite to be your own.' This is what the Lord says: 'Out of your own household I am going to bring calamity on you. Before your very eyes I will take your wives and give them to one who is close to you, and he will sleep with your wives in broad daylight. You did it in secret, but I will do this thing in broad daylight before all Israel.' Then David said to Nathan, 'I have sinned against the Lord.' Nathan replied, 'The Lord has taken away your sin. You are not going to die. But because by doing this you have shown utter contempt for the Lord, the son born to you will die.'"

After Nathan departed, the Lord struck the child that Uriah's wife had borne to David, and he became ill. David pleaded with God for the child. He fasted and spent the nights lying in sackcloth on the ground. The elders of his household stood beside him to get him up from the ground, but he refused and would not eat any food with them. On the seventh day, the child died.

Among the other things that Nathan prophesied to David were: "...the sword will never leave your house..." and "...out of your household I am going to bring calamity..." These things also came to pass just as the prophet had foretold. The fulfillment of these prophecies follows: David's son Amnon rapes his half-sister, Tamar. Tamar's brother, Absalom, then orders his servants to kill Amnon. Absalom then flees to Geshur for three years before returning to Jerusalem.

In 2 Samuel, chapter 16, we learn that David's son Absalom is declared king and seeks David's life, so he flees the city with his men. One of Absalom's advisers, Ahithophel, suggests that he should lie with his father's concubines. They pitch a tent on the roof of the palace so that all of Israel can see. Absalom then pursues David with his army. David instructs his men not to kill his son should the opportunity arise during the conflict. Despite his command, the commander of David's men comes across Absalom hanging in a tree and slays him.

Clearly, the words of Nathan the prophet were divinely inspired. How else could Nathan have known what King David did in secret and what God's punishment would be? Nathan confirmed the source of his inspiration when he said, "This is what the Lord, the God of Israel, says." God communicated to Nathan precisely how David had sinned and what his punishment would be.

Without question, the Bible has qualities unlike any other book ever written. It has authenticated itself as divinely inspired through fulfilled prophecy. The evidence is absolutely compelling that this book was not the product of the minds of the 40 men who penned the words but rather the revelation of a divine being who sees all, including our future.

– 22 –

Dreams and Visions

Another way God communicates with mankind is through dreams and visions. In the second chapter of the Book of Daniel, there is an account of King Nebuchadnezzar having a dream that troubled him. He summoned his magicians, enchanters, sorcerers, and astrologers and asked them to tell him the meaning of the dream. In Daniel 2:4-5 (NIV), we read, "Then the astrologers answered the king in Aramaic, 'O king live forever! Tell your servants the dream, and we will interpret it.' The king replied to the astrologers, 'This is what I have firmly decided: If you do not tell me what my dream was and interpret it, I will have you cut into pieces and your houses turned into piles of rubble.'" When Daniel was told what the king had decreed, he asked his friends Hananiah, Mishael, and Azariah to plead with God for mercy concerning the mystery of the dream so that they would not be executed along with the king's wise men.

That night the mystery was revealed to Daniel in a vision. Daniel then gave thanks to the Lord. Daniel 2:20-23 (NIV) records his comments: "Praise be to the name of God forever and ever; wisdom and power are his. He changes times and seasons; he sets up kings and deposes them. He gives wisdom to the wise, knowledge to the discerning. He reveals deep and hidden things; he knows what lies in darkness, and light dwells with him. I thank and praise you, O God of my fathers: You have given me wisdom and power, you have made known to me what we asked of you, you have made known to us the dream of the king." We then read in Daniel 2:24 that Daniel went to the commander of the king's guard and said to him, "Do not execute the wise men of Babylon. Take me to the king, and I will interpret his dream for him."

The following passages from Daniel 2:26b-43 (NIV) record the king's question to Daniel and his response: "Are you able to tell me what I saw in my dream and interpret it? Daniel replied, 'No wise man, enchanter, magician or diviner can explain to the king the mystery he has asked about, but there is a God in heaven who reveals mysteries. He has shown King Nebuchadnezzar what will happen in days to come. Your dream and the visions that passed through your mind as you lay on your bed are these: As you were lying there, O king, your mind turned to things to come, and the revealer of mysteries showed you what is going to happen. As for me, this mystery has been revealed to me, not because I have greater wisdom than other living men, but so that you, O king, may know the interpretation and that you may understand what went through your mind. You looked, O king, and there before you stood a large statue – an enormous, dazzling statue, awesome in appearance. The head of the statue was made of pure gold, its chest and arms of silver, its belly and thighs of bronze, its legs of iron, its feet partly of iron and partly of baked clay. While you were watching, a rock was cut out, but not by human hands. It struck the statue on the feet of iron and clay and smashed them. Then the iron, the clay, the bronze, the silver and the gold were broken to pieces at the same time and became like chaff on the threshing floor in the summer. The wind swept them away without leaving a trace. But the rock that struck the statue became a huge mountain and filled the whole earth. This was the dream and now we will interpret it to the king. You, O king, are the king of kings. The God of heaven has given you dominion and power and might and glory; in your hands he has placed mankind and the beasts of the field and the birds of the air. Wherever they live, he has made you ruler over them all. You are the head of gold. After you, another kingdom will rise, inferior to yours. Next, a third kingdom, one of bronze, will rule over the whole earth. Finally, there will be a fourth kingdom, strong as iron – for iron breaks and smashes everything – and as iron breaks things into pieces, so it will crush and break all the others. Just as you saw that the feet and toes were partly of baked clay and partly of iron, so this will be a divided kingdom; yet it will have some of the strength of iron in it, even as you saw iron mixed with clay. As the toes were partly iron and partly clay, so this kingdom will be partly strong and partly brittle. And just as you saw the people will be a mixture and will not remain united, any more than iron mixes with clay.'" Daniel 2:46-47 (NIV) goes on to say, "Then King Nebuchadnezzar fell prostrate before Daniel and paid him honor and ordered that an offering and incense be presented to him. The king said to Daniel, 'Surely your God is the God of gods and the LORD of kings and a revealer of mysteries, for you were able to reveal this mystery.'"

How does science explain how Daniel was not only able to interpret King Nebuchadnezzar's dream but also tell the king what he had dreamed? From what source did King Nebuchadnezzar and Daniel receive the information on how the next 2,500 years of human history would unfold? Historians and Bible scholars believe that the head of gold in the king's dream represented the Babylonian kingdom, which King Nebuchadnezzar ruled over. Following his kingdom would come the Medo-Persian empire represented by the arms and chest of silver. After that would come the kingdom of Greece, ruled by Alexander the Great and represented by the belly and thighs of bronze. The fourth kingdom, represented by the legs of iron, was the Roman Empire. The fifth kingdom, represented by the feet of iron and clay, will be a divided kingdom. Following this kingdom will be the final kingdom, the one that will be established when Jesus Christ returns as King of kings.

Daniel 2:44-45 (NIV) describes the final kingdom like this, "In the time of those kings, the God of heaven will set up a kingdom that will never be destroyed, nor will it be left to another people. It will crush all those kingdoms and bring them to an end, but it will endure forever. This is the meaning of the vision of the rock cut out of a mountain, but not by human hands – a rock that broke the iron, bronze, clay, silver, and gold to pieces. The great God has shown the king what will take place in the future. The dream is true, and the interpretation is trustworthy."

The kingdom represented by the feet of iron and clay is thought to be the antichrist kingdom that will be smashed by the final king, known as the King of kings and Lord of lords. He will depose all that is wicked and overcome the kingdom of the antichrist.

If Daniel was able to tell the king what he dreamed and accurately interpret the dream, is it not certain that his prophecy regarding the final kingdom will also come to pass?

Another example of how God communicated the future in a dream is found in the story of Joseph, the son of Jacob. The revelation revealed what was about to happen to Jacob's family. It was communicated through a dream given to Jacob's son Joseph when he was a young boy. Genesis 37:5-11 (NIV) describes the story this way, "Joseph had a dream, and when he told it to his brothers, they hated him all the more. He said to them, listen to this dream I had: We were binding sheaves of grain out in the field when suddenly my sheaf rose and stood upright, while your sheaves gathered around mine and bowed down to it. His brothers said to him, do you intend to reign over us? Will you actually rule us? And they hated him all the more because of his dream and what he had said. Then he had another dream, and he told it to his brothers. Listen, he said, I had another dream, and this time the sun and the moon and eleven stars were bowing down to me. When he told his father as well as his brothers, his father rebuked him and said, what is the dream you had? Will your mother and I and your brothers actually come and bow down to the ground before you?"

Because of their hatred for Joseph, his brothers sold him into slavery, and he was moved to Egypt, where he was later imprisoned. In prison, he gained favor with the captain of Pharaoh's guard and was put in charge of managing the prison on behalf of the captain. While he was in prison, Pharaoh became angry with his cupbearer and his baker and had them both thrown into prison with Joseph. One night they both had dreams which troubled them. Genesis, 40:6-8 (NIV) says, "When Joseph came to them the next morning, he saw that they were dejected. So he asked Pharaoh's officials who were in custody with him in his master's house, why are your faces so sad today? We both had dreams, they answered, but there is no one to interpret them. Then Joseph said to them, do not interpretations belong to God? Tell me your dreams." Note that Joseph did not suggest that he could interpret their dreams but that dream interpretations belonged to God.

Genesis 40:9-13 (NIV) goes on to say, "So the chief cupbearer told Joseph his dream. He said to him, 'In my dream I saw a vine in front of me, and on the vine were three branches. As soon as it budded, it blossomed, and its clusters ripened into grapes. Pharaoh's cup was in my hand, and I took the grapes, squeezed them into Pharaoh's cup and put the cup in his hand.' This is what it means, Joseph said to him. 'The three branches are three days. Within three days Pharaoh will lift up your head and restore you to your position, and you will put Pharaoh's cup in his hand, just as you used to do when you were his cupbearer.'"

The story continues in Genesis 40:16-22 (NIV), "When the chief baker saw that Joseph had given a favorable interpretation, he said to Joseph, 'I too had a dream: On my head were three baskets of bread. In the top basket were all kinds of baked goods for Pharaoh, but the birds were eating them out of the basket on my head.' This is what it means, Joseph said. 'The three baskets are three days. Within three days Pharaoh will lift off your head and hang you on a tree. And the birds will eat away your flesh.' Now the third day was Pharaoh's birthday, and he gave a feast for all his officials. He lifted up the heads of the chief cupbearer and the chief baker in the presence of his officials: He restored the chief cupbearer to his position, so that he once again put the cup into Pharaoh's hand, but he hanged the chief baker, just as Joseph had said to them in his interpretation."

Another example of a vision given in a dream to the King of Egypt (Pharaoh) is recorded in the 41st chapter of Genesis. It occurred during the time of the Jewish captivity. Genesis 41:1-7 (NIV) states, "...Pharaoh had a dream: He was standing by the Nile, when out of the river there came up seven cows, sleek and fat and they grazed among the reeds. After them, seven other cows ugly and gaunt, came up out of the Nile and stood beside those on the riverbank. And the cows that were ugly and gaunt ate up the seven sleek, fat cows. Then Pharaoh woke up. He fell asleep again and had a second dream: Seven heads of grain, healthy and good, were growing on a single stalk. After them, seven other heads of grain sprouted – thin and scorched by the east wind. The thin heads of grain swallowed up the seven healthy, full heads. Then Pharaoh woke up; it had been a dream."

The passage continues in Genesis 41:8-13 (NIV), "In the morning his mind was troubled, so he sent for all the magicians and wise men of Egypt. Pharaoh told them his dreams, but no one could interpret them for him. Then the chief cupbearer said to Pharaoh, 'today I am reminded of my shortcomings. Pharaoh was once angry with his servants, and he imprisoned me and the chief baker in the house of the captain of the guard. Each of us had a dream the same night, and each dream had a meaning of its own. Now a young Hebrew was there with us, a servant of the captain of the guard. We told him our dreams, and he interpreted them for us, giving each man the interpretation of his dream. And things turned out exactly as he interpreted them to us: I was restored to my position, and the other man was hanged.'"

Genesis 41:14-16 (NIV) says, "So Pharaoh sent for Joseph, and he was quickly brought from the dungeon. When he had shaved and changed his clothes, he came before Pharaoh. Pharaoh said to Joseph, 'I had a dream, and no one can interpret it. But I have heard it said of you that when you hear a dream you can interpret it.' 'I cannot do it, Joseph replied to Pharaoh, but God will give Pharaoh the answer he desires.' Then Pharaoh explained his dream to Joseph."

Genesis 41:25-32 (NIV) records Joseph's response, "'The dreams of Pharaoh are one and the same. God has revealed to Pharaoh what he is about to do. The seven good cows are seven years, and the seven good heads of grain are seven years; it is one and the same dream. The seven lean, ugly cows that came up after they did are seven years, and so are the seven worthless heads of grain scorched by the east wind: They are seven years of famine. It is just as I said to Pharaoh: God has shown Pharaoh what he is about to do. Seven years of great abundance are coming throughout the land of Egypt, but seven years of famine will follow them. Then all the abundance in Egypt will be forgotten, and the famine will ravage the land. The abundance in the land will not be remembered, because the famine that follows it will be so severe. The reason the dream was given to Pharaoh in two forms is that the matter has been firmly decided by God, and God will do it soon.'"

Joseph went on to advise Pharaoh in Genesis 41:33-36 (NIV), "'And now let Pharaoh look for a discerning and wise man and put him in charge of the land of Egypt. Let Pharaoh appoint commissioners over the land to take a fifth of the harvest of Egypt during the seven years of abundance. They should collect all the food of these good years that are coming and store up the grain under the authority of Pharaoh, to be kept in the cities for food. This food should be held in reserve for the country, to be used during the seven years of famine that will come upon Egypt, so that the country may not be ruined by the famine.'"

Genesis 41:37-41 (NIV) says, "The plan seemed good to Pharaoh and to all his officials. So Pharaoh asked them,

'Can we find anyone like this man, one in whom is the spirit of God?' Then Pharaoh said to Joseph, 'Since God has made all this known to you, there is no one so discerning and wise as you. You shall be in charge of my palace, and all my people are to submit to your orders. Only with respect to the throne will I be greater than you.' So Pharaoh said to Joseph, 'I hereby put you in charge of the whole land of Egypt.'"

After Joseph was promoted to the number two position in Egypt and the seven years of abundance had ended, the famine came. Genesis 42:1 (NIV) describes the impact on Joseph's family: "When Jacob learned that there was grain in Egypt, he said to his sons, 'Why do you just keep looking at each other?' He continued, 'I have heard that there is grain in Egypt. Go down there and buy some for us, so that we may live and not die.'"

Genesis 42:6b (NIV) says, "So when Joseph's brothers arrived, they bowed down to him with their faces to the ground." This was the fulfillment of Joseph's dream, where he saw his brother's sheaves bowing down to his sheave. Even Joseph's father and mother were subject to his rule. They moved to Egypt to escape the famine, and Joseph gave them one of the choicest places to raise their livestock, in the district of Rameses. He also gave them property. This fulfilled his second dream that his father, mother, and eleven brothers would be subject to him.

Where does knowledge of future events come from, if not from an omnipotent all-knowing God? Can science, materialism, or random chance explain the source of this information? They cannot. These historical events regarding the fulfillment of prophecy can only be attributed to an Almighty God, who sees the end from the beginning and exists outside of time, space, and matter.

- 23 -

Logic and Reason

The Merriam-Webster dictionary defines *logic* as: "a science that deals with the principles and criteria of the validity of inference and demonstration: the science of the formal principles of reasoning." [99] It defines *reason* as: "a rational ground or motive; the thing that makes some fact intelligible; a sufficient ground of explanation or of logical defense *especially* something (such as a principle or law) that supports a conclusion or explains a fact." [100]

There are three fundamental laws of *logic*: the law of contradiction, the law of the excluded middle, and the principle of identity. [101] They are defined as follows: The law of contradiction: for any proposition x, it is impossible for both x and not x to be true. Second, the law of the excluded middle: either x or not x must be true, and there can be no third or middle true proposition. Finally, the law of identity: whatever is, is. If a given proposition is defined as x, then x = x.

Does Darwin's theory of evolution adhere to these three fundamental laws? Does it make sense logically? Darwin's theory suggests that all life was spontaneously generated from non-life, i.e., inorganic matter. The definition of inorganic is: "Of or pertaining to substances that are not of organic origin," i.e., not living. [102] The definition of organic is: "Of, relating to, or derived from living organisms." [103] The first law of logic, the law of contradiction, states that for any proposition x, it is impossible for both x and not x to be true. Therefore, inorganic non-living material can't be both living and non-living. To my knowledge, there are no known documented examples of a non-living substance or material suddenly coming to life. Making an organic compound in a lab is not the same thing as making life. The second law states that either x or not x must be true. Either inorganic material is living, or it isn't living. We know that inorganic material is not living. The third law of logic states that whatever is, is. So, inorganic material is inorganic, it is not alive. Even the definition of organic material suggests it is derived from living organisms. It's not the product of inorganic matter. The idea that non-living, inorganic elements could become living seems implausible based on logic and common-sense reasoning.

In fact, with all the sophistication and intelligence of modern science, there is not a single example of anyone coming remotely close to creating anything resembling a living cell or any other simple life form. Today, we're asked to believe that complex structures like DNA, RNA, proteins, lipids, fatty acids, glycerol, and phosphate, which are necessary for forming even the simplest cell, somehow came together purely by chance, without any intelligence guiding the process or providing the necessary information or instructions. The probability that this could have occurred by chance is less than one in a trillion trillion

trillion trillion trillion trillion trillion trillion...add many more pages of trillions.

However, the world is still holding on to this theory, even as recent discoveries related to the complexity of life make the theory far less plausible than when Darwin first proposed it. With recent discoveries regarding the complexity of cells, proteins, and DNA, the notion that life emerged from nonliving material has become increasingly implausible with many of today's scientists.

At the time Darwin conceived the theory, science thought the cell was a glob of protoplasm and not very complex. We now know that the simplest cells are far more complicated than the largest cities on Earth. Try to imagine a city with the complexity of New York spontaneously coming into existence without an intelligent designer. It's unimaginable.

What about the Big Bang Theory? How well does it stand up to the laws of logic? The first law, the law of contradiction, states that for any proposition x, it is impossible for both x and not x to be true. The Big Bang Theory is based on the premise that nothing exploded and became something—the universe. The first law of logic states that x (nothing) and y (something) can't both be true. If it is true that the Big Bang started with nothing, it can't be true that it became something when nothing expanded.

How does it stand up to the second law, the law of the excluded middle, which says that either x or not x must be true? If x is true, that the Big Bang resulted when nothing expanded, then not x, something (our material universe), can't be true.

Lastly, how does the Big Bang Theory hold up against the third law of logic, the law of identity, which states that whatever is, is? If a given proposition is defined as x (nothing), then x = x, i.e., nothing = nothing. Therefore, x (nothing) cannot be y (something) based on the laws of logic. You don't get something when nothing expands, no matter how fast the rate of expansion. If "is" equals nothing, then that is what it remains, nothing, based on the third law of logic.

Evolution and the Big Bang defy the fundamental laws of logic, and a growing mountain of evidence opposing these theories is beginning to undermine the credibility of both.

- 24 -

The Laws of Thermodynamics

Regarding the laws of thermodynamics, Albert Einstein had this to say: "It is the only physical theory of universal content, which I am convinced, that within the framework of applicability of its basic concepts will never be overthrown." [104] This chapter contrasts the laws of thermodynamics with evolution and the Big Bang Theory.

The first and second laws of thermodynamics relate to energy and they are two of the most fundamental scientific laws governing our universe. Everything in our universe exists in the form of energy, and every action is the result of some form of energy conversion. Therefore, to understand our universe, one must understand the laws of thermodynamics. "The first law, also known as the Law of Conservation of Energy, states that energy can't be created or destroyed, but only transformed or transferred." [105] Because energy and matter don't create themselves, they must be the product of an outside force or actor unless they have eternally existed. However, current science believes the universe had a beginning and that energy and matter were created from nothing. This violates the first law of thermodynamics, which states that energy cannot be created or destroyed.

The second law relates to entropy. Merriam-webster.com defines *entropy* as: "a measure of the unavailable energy in a closed thermodynamic system that is also usually considered to be a measure of the system's disorder…, the degradation of the matter and energy in the universe to an ultimate state of inert uniformity; *entropy* is the general trend of the universe toward death and disorder. – James R. Newman." [106] The Law of Entropy states that everything in the universe tends toward randomness and disorder over time. That is, things deteriorate, break down, and wear out. However, the Big Bang and evolution both depend on an upward progression, both in order and complexity—the opposite of the second law of thermodynamics.

We're told that an expansion of nothing created the universe, our solar system, the laws of quantum physics, the cosmological constants, and dozens of other anthropic principles that enable life to exist. We're asked to believe that on a primordial Earth, through a process called abiogenesis, inorganic materials self-organized into an incomprehensibly complex arrangement of compounds and miraculously sprang to life.

The laws of thermodynamics stand in direct opposition to both the theory of evolution and the Big Bang. If energy can't be created or destroyed, what is the source of its origin? If the Law of Entropy is true, which states that all things deteriorate and become less orderly over time, how did life evolve into more complex organisms starting with inorganic material? There is no evidence or detailed description of how inorganic material evolved into living organisms. The fossil record is completely void of the transitional life forms that Charles Darwin said would be present if his theory were true. Both evolution and the Big Bang theory break down rapidly when confronted with evidence!

The newest evidence comes from data collected by the James Webb Space Telescope (JWST), launched on Christmas Day of 2021. This data contradicts some of the most fundamental ideas postulated by the Big Bang Theory. For example, some of the data suggests there are galaxies a hundred times bigger than our Milky Way Galaxy that appear to have formed in just half a billion years. This should not be possible based on current viewpoints espoused by the Big Bang Theory. [107]

Everything we know, from the subatomic realm to the outer reaches of space, exhibits a level of complexity that is beyond our imagination. Neither of these theories passes the test of logic and reasoning, and they both defy the laws of thermodynamics.

– 25 –

Our Fine-Tuned Universe

Discoveries made by the James Webb Telescope may be rewriting the timeline for the universe's beginning. Our modern science texts claim that 13.8 billion years ago, nothing exploded and created everything we see in the universe. As material and energy released by the Big Bang began to accelerate across space, it miraculously started to organize into near-perfect spherical objects. These spherical objects miraculously ordered themselves into solar systems and galaxies as they raced through space. The majority of these (moons, planets, suns, solar systems, and galaxies) rotate in a counterclockwise direction due to the law of angular momentum and the rapid expansion of energy that resulted from the Big Bang. [108]

But surprisingly, Venus rotates in a clockwise direction, as do some of the moons in our solar system. Even more peculiar is that the planet Uranus rotates on its side at a 90-degree angle to its orbit. Why examples exist of objects rotating in a direction contrary to the law of angular momentum is puzzling. In an expanding universe, why do some galaxies appear to be moving toward us rather than away from us, as suggested by their blue light shifts? Of the hundreds of billions of galaxies in our known universe, only around a hundred appear to have a blue light shift, while all others have red light shifts, suggesting they are moving away from us.

Imagine if we could travel back to the beginning of our universe, which was hypothesized to have occurred 13.8 billion years ago, and make predictions about what would happen after the initial Big Bang. How many of us would have been able to predict that the event would lead to the creation of our current universe? Who could have foreseen that from the expansion of nothing hundreds of billions of galaxies would emerge, each containing hundreds of billions of stars, planets, and moons? How many could have predicted that only one planet out of the hundreds of trillions of planets would have the exact conditions for life to occur? Who would have predicted that our unique planet would be the only one in the universe that we're aware of, with just the right amount of carbon, oxygen, nitrogen, and hydrogen to support all the life forms we observe on Earth? Who would have predicted the formation of our solar system? Who could have predicted that the Earth would be positioned at exactly the right distance from the Sun to make life possible? How many would have predicted that the speed of Earth's rotation would be exactly right for sustaining life? Who would have predicted that the tilt of the Earth's axis would be precisely what was needed to provide the Earth with four seasons? Who would have predicted that the planets in our solar system would form with just the right masses and a perfect balance of gravitational forces and orbital dynamics to maintain stable orbits for thousands of years?

To have predicted the outcome of the universe we now observe as a result of a Big Bang would have been virtually impossible. The astronomical odds of getting the universe we observe are beyond our understanding. When we calculate the probability that it could have occurred by chance, we must conclude that it wasn't possible, at least not by random chance. There is too much order, too much precision, too much information, and too much design for it to have been anything other than a supernatural creation.

In their book, *I Don't Have Enough Faith to Be an Atheist*, Norman L. Geisler and Frank Turek described it this way: "Scientists are now finding that the universe in which we live is like a diamond-studded Rolex, except the universe is even more precisely designed than the watch. In fact, the universe is specifically tweaked to enable life on Earth. A planet with scores of improbable and interdependent life-supporting conditions that make it a tiny oasis in a vast and hostile universe." They go on to say, "The extent of the universe's fine-tuning makes the anthropic principle perhaps the most powerful argument for the existence of God." [109]

What is the anthropic principle, and why is it such a compelling argument for the existence of God? The definition of anthropic is: "relating to human beings or their existence," and "principle means law." [110] So, the anthropic principle has to do with the fine-tuning of the universe to support human existence.

It is well known that our existence in this universe depends on numerous cosmological constants and parameters whose values must fall within a very narrow range. If even a single variable were off, even slightly, life could not exist. The extreme improbability that these variables would align so propitiously in our favor merely by chance has led some scientists and philosophers to propose that it was God who providentially engineered the universe to enable life. In defining the anthropic principle, the Encyclopedia Britannica suggests that the universe appears to have been fine-tuned for our existence. [111]

Consider the following facts related to the speed at which the Earth, our solar system, and our galaxy are moving through space and whether these movements are the result of a primordial Big Bang or a master design. The Earth rotates on its axis at a speed of more than 1,000 miles per hour. We are orbiting around the sun at a speed of more than 66,000 miles per hour. Our sun and solar system orbit the Milky Way galaxy at a speed of more than 400,000 miles per hour, and our Milky Way galaxy is moving through space at more than 1.3 million miles per hour. [112]

Now consider the fact that the Earth has been flying through space at more than 1,700 times the speed of sound for over 4.5 billion years [the estimated age of the Earth, according to science], and we haven't run into any significant debris from that initial expansion that could have destroyed our planet. If we had, we wouldn't be here. Neither the Sun nor any of the planets in our solar system has run into any sizeable debris that could have destroyed them. Nor have we passed too close to a black hole that would have ended our planet's existence. That's some kind of extraordinary luck: 14 billion years without a major event that would have ended our existence.

But maybe it wasn't luck. Maybe the whole universe had a designer, and everything was set in motion with a precision that would make atomic clocks look primitive. Consider the fact that if the Earth and the other planets in our solar system were orbiting the Sun a little faster, they would fly off into space. Or if they were orbiting a little slower, they would be pulled into the Sun by its strong gravitational field. Everything seems to be moving at just the right speed to prevent utter disaster. We're either lucky beyond imagination, or there is a divine being controlling every detail of our universe, from the subatomic quantum realms to the farthest regions of space, from the speed of electrons to the speed of planetary orbits, from the internal workings of a human cell to the cosmological constants that enable everything to operate in perfect harmony.

More and more scientists are now suggesting that our universe appears to be remarkably fine-tuned to support life. But just how precise does the fine-tuning have to be to provide the conditions needed to support life? Oxford University Professor of Mathematics John Lennox, in an essay titled "Is There a God (What is the Chance the World is the Result of Chance?)" quotes Roger Penrose, a renowned Oxford University mathematical physicist who calculated the probability that our universe could have come into existence by chance as one chance in 10 to the power of 10 to the power of 123 (10^10^123). [113] As Penrose put it, that is a "number which would be impossible to write out in the usual decimal way because even if you were able to put a zero on every particle in the universe, there would not be enough particles to do the job." [114]

Astrophysicist Hugh Ross has also calculated the probability that 122 constants, which allow our universe to exist in the precise balance needed to support life came about by chance is one chance in 10 to the power of 138 (10^138). [115] The number of atoms in the entire universe is between 10 to the power of 70 and 10 to the power of 80. Another way to express Ross's probability of getting our 122 constants by chance is less than the probability of selecting one marked atom from a trillion, trillion, trillion, trillion, trillion universes, each with the same number of atoms as our own universe.

When evaluating the likelihood of our universe arising by random chance, the probabilities presented by thinkers like Roger Penrose and Hugh Ross highlight just how implausible this scenario seems. This may explain why some atheists are drawn to the multiverse hypothesis. Advocates of this idea propose that an infinite number of universes could exist alongside our own. Under this hypothesis, the extreme improbability of our universe being a random occurrence is mitigated by the notion that, with an infinite number of opportunities, even the most unlikely events become possible. In essence, proponents argue that no matter how remote the odds, any outcome can appear plausible when the denominator is infinity. The only problem with this hypothesis is that there is no definitive evidence that another universe exists besides our own, much less an infinity of universes.

- 26 -

The Random Chance Hypothesis

The prevailing view of how our universe came into existence is that there was a Big Bang that produced the material and energy needed to form the cosmos. Subsequently, life emerged spontaneously from nonliving matter. It all happened without the aid of intelligent input, design, or information to direct the processes that brought it about. But how plausible are these theories?

According to the OpenAI source ChatGPT, the Hebrew Old Testament has 304,805 letters. The Greek New Testament has approximately 838,000 letters. The probability of randomly selecting all the letters needed to reproduce both the Old and New Testaments in the exact right order is 10^-6,138,928.44, effectively zero, according to ChatGPT. [116] Could an infinite number of random drawings of Hebrew and Greek letters eventually produce the Old and New Testaments, word for word, letter for letter?

Science suggests that the intricate fine-tuning observed in the universe, e.g., living organisms, cellular structures, and DNA, all came about by chance over billions of years. Yet, the odds against any of these occurrences happening by random chance are astronomically greater than the odds of blindly drawing the correct sequence of letters to produce the Old and New Testaments. If the latter seems impossible, why is it not reasonable to question whether the complexity we see in our universe could be the result of random chance?

In previous discussions regarding the fine-tuning of the universe and the anthropic principle, some incredibly large numbers were explored. One of these numbers was related to the probability of getting, by chance, 122 constants needed to support life. One of those constants is the force of gravity. According to Brandon Carter, if it had been stronger or weaker by 1 part in 10^40, life-sustaining stars like the sun could not exist, which means life would not exist. [117] To put this into perspective, if the fine-tuning of gravity were placed on a ruler that stretched across the entire universe we wouldn't be able to move its position one millionth of an inch without eliminating any possibility for life.

If you are questioning whether this ruler example is a fair representation of how little gravity would have to change for our universe not to exist, let's compare the two numbers. The current estimate of the distance across the universe is 93 billion light-years from end to end. One light-year is approximately 5.9 trillion miles. The distance across the universe in miles is calculated by multiplying 93 billion by 5.9 trillion, which gives us 548.7 sextillion miles.

93,000,000,000 (light years)
<u>   x 5,900,000,000,000 (miles in one light year)</u>
548,700,000,000,000,000,000,000 (distance in miles)

By multiplying the feet in a mile (5,280) by the number of inches in a foot (12), we get the total number of inches in a mile (63,360). The number of inches across the universe is calculated by multiplying 548.7 sextillion miles by 63,360 inches. The distance is 34.8 octillion inches.

548,700,000,000,000,000,000,000
$$\underline{ \times 63,360}$$
34,765,600,000,000,000,000,000,000,000 (total inches)

For a ruler that stretches across the entire span of the universe to be as fine tuned as gravity, you'd have reduce the scale on the ruler from inches to 10, billionths of inches.

Inches Across the Universe
Compared to
The Fine-Tuning of Gravity

34765600000000000000000000000
– distance in inches across the universe

.0000000000000000000000000000000000001
– fine-tuning of gravity

Note that there are 11 more digits in the fine-tuning of gravity than the number of inches across the universe. That means gravity is *10 billion times* more fine-tuned than a ruler that stretches across the universe measured in inches! This suggests that if we placed gravity at its current strength somewhere on that ruler, we could not move it

1/1,000,000 of an inch, or we wouldn't be here; neither would our universe as we know it.

Is it plausible to believe that a Big Bang created this incredibly precise gravitational force and more than a hundred other finely tuned laws and constants that enable our universe to support life? As improbable as it is that gravity would be fine-tuned to the 40th decimal place by random chance, consider how much more unlikely the probability of human beings arising by random chance is. New scientific discoveries, especially those related to mapping the human genome, have not been helpful in advancing the theory of evolution.

The human body has approximately 30 trillion cells. [118] Each of these cells contains DNA, which has the instructions for building proteins that are vital to the functioning of the cells. Within the human body there are approximately 20,000 unique protein-encoding genes responsible for more than 100,000 unique proteins. [119] These proteins catalyze chemical reactions, synthesize and repair DNA, transport materials across the cell, receive and send chemical signals, respond to stimuli, and provide structural support. [120] Is it reasonable to believe that the instructions and information for building cells, DNA, and 100,000 unique proteins came into existence by random chance?

Consider the fact that proteins are made up of 20 different amino acids, which form protein chains that are typically about 300 residues long. Each of the amino acids in the chain must be in the right order for the protein to function properly. Considering that each of the 30 trillion cells in the human body comprises thousands of proteins, the complexity of each cell is staggering. "The largest known proteins are the titins, a component of the muscle sarcomere, with a molecular mass of almost 3,000 kDa and a total length of almost 27,000 amino acids." [121] Some estimates place the number of amino acids in the titin protein as high as 34,000. [122] Imagine the odds of getting 20 different amino acids in a titin protein in the exact order by random chance. The odds would be 20 to the power of 27,000, ($20^{27,000}$) or 20 to the power of 34,000 ($20^{34,000}$) on the high side! Now, imagine the probability of getting a human being by random chance, with 30 trillion cells all working together to perform the various bodily functions. Then, consider that each individual cell has thousands of unique proteins, all performing their assigned functions. The odds of all this happening by random chance are beyond comprehension! To believe that it happened by chance requires a faith that contradicts mathematical reason.

- 27 -

The Anthropic Principle

New discoveries are beginning to persuade some scientists that the exceptionally fine-tuned cosmological constants that enable life on Earth appear to be the handiwork of an intelligent designer. Whether examining the outer reaches of our universe or the inner workings of our own solar system, the precision behind the fine-tuning appears to be specifically designed to support life. The likelihood this fine-tuning occurred by chance is not just highly unlikely, many scientists have deemed it impossible.

An article in U.S. News & World Report states, "So far, no theory is even close to explaining why physical laws exist, much less why they take the form they do. Standard Big Bang Theory, for example, essentially explains the propitious universe in this way: 'Well, we got lucky.'" [123] When considering whether we just got lucky or whether the precision we're seeing is the result of an intelligent designer, let's examine why some scientists are baffled by the new evidence.

In his January 1, 2002, article, "Anthropic Principle: A Precise Plan for Humanity," Hugh Ross states, "In 1961, astronomers acknowledged just two characteristics of the universe as 'fine-tuned' to make physical life possible. The more obvious one was the ratio of the gravitational force constant to the electromagnetic force constant. It cannot differ from its value by any more than one part in 10 to the power of 40 (that's one part in ten thousand, trillion, trillion, trillion), without eliminating the possibility for life." [124]

Over the past sixty years, scientists have identified many more constants, known as anthropic principles or extreme fine-tuning in physics, quantum mechanics, the strong and weak nuclear forces, electromagnetism, and many other constants, both dimensional and nondimensional, that allow life to exist on our planet. Modern science does not understand where these astonishingly precise principles, laws, and finely tuned constants came from. The idea that they all could have occurred by random chance or coincidence is considered mathematically impossible. According to Borel's Law of Probability, any number that exceeds 10 to the power of 50 is impossible. [125]

Below is a list of some of the finely tuned parameters that enable life to exist on our planet, any one of which, if it were off even slightly, would end the possibility of life on Earth.

1. The Expansion Rate of the Universe: According to modern science, if the universe had expanded more or less rapidly than it did, we would not be here. Either the planets would not have formed (if the expansion had been faster), or the universe would have collapsed on itself (if the expansion had been slower).

2. Speed of Light: The laws of physics can be described as a function of the speed of light. Even a slight change in the speed of light would alter the other constants and eliminate the possibility of life on Earth.

3. Oxygen: If the oxygen level were a few percentage points higher, fires would erupt spontaneously. If it were a few percentage points lower, we would suffocate.

4. Atmospheric Transparency: If our atmosphere were less or more transparent, we would receive either too little or too much solar radiation to sustain life.

5. Water Vapor Levels: If water vapor levels were higher, we would experience a runaway greenhouse effect, and atmospheric temperatures would rise to a level too high to sustain human life. If water vapor levels were lower, the Earth would be too cold to support human life.

6. Carbon Dioxide: If CO2 levels were slightly higher, the Earth's temperature would be too hot; if slightly lower, it would affect the photosynthesis of plants, and we could starve or suffocate.

7. Jupiter: If Jupiter were not in our solar system and in its current orbit, the Earth would be vulnerable to a barrage of asteroids, meteors, and comets. The

gravitational field of Jupiter protects the Earth from cosmic debris.

8. Earth's Crust: The thickness of the Earth's crust determines how much oxygen is transferred to it. If it were thicker, too much oxygen would be transferred to the crust, making life improbable. If the crust were thinner, we would experience too much volcanic and tectonic activity to support life.

9. Earth's Rotation: If the Earth's rotation were shorter, wind velocities would increase greatly; if it were longer, day and night temperature differences would be too great, impacting food production.

10. Earth's Axis Tilt: A slight variation in the current 23-degree axial tilt would alter the Earth's surface temperatures, negatively affecting the sustainability of life. [126]

Discovery Institute published an article titled "The Top Six Lines of Evidence for Intelligent Design" in February 2021. Their second line of evidence was related to the fine-tuning of the universe. The article states, "The finely tuned laws and constants of the universe are an example of specified complexity in nature. They are complex in that their values and settings are highly unlikely." [127] The following probabilities demonstrate the level of fine-tuning and can be interpreted as the likelihood of these constants arising by chance.

1. Gravitational constant: The probability that the strength of gravity is exactly calibrated to allow for our solar system and universe by chance is one in 10^40, or one in 1,000,000,000,000,000, 000,000,000,000,000,000,000,000. [128]

2. Electromagnetic force versus the force of gravity: The probability that these two forces would be perfectly calibrated to allow for our solar system and universe to exist in its present state is one in 10^37. [129]

3. Ratio of electrons to protons: One in 10^37. [130]

4. Expansion rate of the universe: One in 10^55 (add 15 more zeros to the gravitational constant number). [131]

5. Mass density of universe: One in 10^59 (add 19 more zeros to the gravitational constant number). [132]

6. Cosmological constant: One in 10^120 (add 80 zeros to the gravitational constant number. [133]

7. Distribution of mass-energy: The probability that chance was responsible for the perfect conditions that existed at the beginning or initial state of the universe, i.e., the distribution of

mass-energy, or extremely low entropy, is astronomically small. "In *The Road to Reality*, physicist Roger Penrose estimates that the odds of the initial low entropy state of our universe occurring by chance are on the order of one in 10^10^123." [134] Add trillions more zeros than there are atoms in the universe to the gravitational constant number!

Let's compare the probability of getting the level of fine-tuning needed to support the initial formation of the universe, i.e., the distribution of mass-energy just after the Big Bang, with the probability of winning the Powerball lottery. According to cbsnews.com, "The odds of winning the Powerball jackpot are one in 292 million." [135] The probability that random chance produced the fine-tuning level seen in the mass-energy distribution at the beginning of the universe is estimated to be one chance in 10^10^123. By comparison, you would have to win the Powerball jackpot 1.181 x 10^122 consecutive times to equal the probability of 10^10^123. [136]

Here is another example of how ChatGPT described the enormity of the number 10^10^123. First, it pointed out that the empty space in our universe constitutes 99.9999999% of the total space. It then said if you wrote this number in standard form and the zeros were the size of an atom, the entire observable universe isn't large enough to contain all the zeros. [137] Try to imagine a number so big that if you filled all the space in the universe with zeros the size of atoms, there wouldn't be enough

174

space. Then ask yourself whether it is more likely that this universe was a chance coincidence or whether the incomprehensible precision we observe is more likely the hand of an omnipotent God.

If you understand the math, it's virtually impossible to believe this universe is the result of a random chance expansion of nothing into something, the theory behind the Big Bang. To put one's faith in probabilities of this magnitude not only defies common sense and reason, it denies the clear evidence that God is the Creator. That's why He says in Romans 1:20, NIV, "For since the creation of the world, God's invisible qualities – his eternal power and divine nature – have been clearly seen, being understood from what has been made, so that men are without excuse."

God designed the universe as a life-sustaining home for His creation, a world that was both breathtakingly beautiful and intricately ordered. He created humanity in His image, desiring a deep and meaningful relationship with us. However, our sins fractured that relationship. To restore it, God sent His Son to bear the penalty for our sins, enduring the agony of the cross on our behalf.

We are now faced with a choice: to accept His gracious gift of salvation by trusting in Jesus' sacrificial death and surrendering to His Lordship or to reject it and follow our own path. God has provided abundant evidence of His existence, the truth of His Word, and His love for us. The choice—and its eternal consequences—now rests with each of us.

– 28 –

Master Designer or Evolution

The more science learns about the quantum realm, the human genome, the inner workings of the cell, and the extremely fine-tuned universe, the more evidence we accumulate for the necessity of a master designer. When Charles Darwin released his famous book, *On The Origin of Species*, science was primitive compared to today's standards. Virtually nothing was understood regarding the inner workings of the cell. In fact, the functions of DNA, RNA, and proteins were unknown to Darwin's generation.

The human genome contains tens of thousands of protein-coding genes, and individual cells house millions of protein molecules. Charles Darwin and the scientists of his time were unaware of the immense complexity of human cells or the vast information-carrying capacity of DNA. Given their limited understanding of cells, it seemed plausible to Darwin that the microevolution (small changes within a kind) that he observed within a species could explain how life might have evolved.

What Darwin observed in the finch population on the Galapagos Islands had nothing to do with macroevolution (the idea that one species or kind could evolve into a completely different kind). What Darwin observed within the finch population was microevolution, which is the selection of different characteristics already existing within the finch gene pool. However, after observing minor changes within the finch population, Darwin extrapolated his observations from the micro level to the macro level. He suggested that over a long period of time, the small changes he observed in finch beak sizes might eventually lead to a totally different species or kind.

Given recent scientific discoveries in genetics, including the self-correcting function in DNA replication and the rare occurrence of mutations, Darwin's theory not only seems implausible, but appears impossible. But in spite of this new evidence, the majority of academia has been reluctant to relinquish the theory. The most likely reason is that the rational alternative to Darwin's theory is the one presented in Genesis 1:21 (NIV), where we read, "So God created the great creatures of the sea and every living and moving thing with which the water teems, according to their kinds, and every winged bird according to its kind. And God saw that it was good." Darwin's theory of evolution, which is based on random chance, long periods of time, mutations, and survival of the fittest, is now being doubted even by some of its staunchest supporters. This is because new scientific discoveries present increasingly difficult problems for the theory to explain.

It is well established that science has never observed macroevolution, which appears to contradict everything we now know about genetic biology. Darwin believed that over time, his theory would be validated by the discovery of millions of transitional life forms, which he thought would eventually appear in the fossil record. The transitional life forms, which he assumed would be abundant, have not been found. Darwin later wrote in Volume 2, Chapter 9, 6th edition of his book, *On The Origin of Species,* "The absence of transitional forms between species...presses hardly on my theory, and ...Geology assuredly does not reveal any such finely graded organic chain; this is the most obvious and gravest objection urged against my theory."

It has been more than 150 years since Darwin wrote those words, and we're still looking for those transitional life forms. So, where are they? If the theory is true, hundreds of millions of fossils should show the links between the various species and kinds as they evolved over time, just as Darwin predicted. However, if the theory is false, you would not expect to find transitional life forms, which is exactly what the fossil record shows.

The Bible says in Genesis 1:11 (NIV), "Then God said, 'Let the land produce vegetation: seed-bearing plants and trees on the land that bear fruit with seed in it, *according to their various kinds.*" God spoke these things into existence on the third day. Genesis 1:26 (NIV) states that on the sixth day, God said, "Let us make man in our image, in our likeness, and let them rule over the fish of the sea and the birds of the air, over the livestock, over all the earth, and over all the creatures that move along the ground." The biblical account of creation aligns with the fossil record and biological reproductive processes for all known life forms.

The entire body of scientific evidence on reproduction aligns with the biblical account, which states that all life forms reproduce after their own kind. I am aware of no credible scientific evidence that links two different species or kinds through an evolutionary process. In addition, science has been unable to explain how the vastly complex information and instructions stored in DNA could have been produced through an evolutionary process. Nor does the theory of evolution provide any explanation for how the first life on Earth could have spontaneously generated from non-organic matter. There is simply no proof that macroevolution between kinds has ever occurred.

At the same time that Darwin was formulating his theory of evolution, Gregor Mendel, the Father of Modern Genetics, was developing his genetic laws. Mendel's genetic laws were based on his observations of how the principles of heredity affected subsequent generations of pea plants. Many scholars have speculated that if Mendel's genetic laws had received the same level of attention as Darwin's theory of evolution, Darwin's theory would never have gained widespread support. This is because Mendel's genetic laws align with the creation account in Genesis and with observed science. They suggest that all life reproduces after its own kind. We now know that the characteristics of the offspring of all life forms are subject to what's available in the gene pool for that species. Mendel's laws of genetics are a major stumbling block for the theory of evolution.

We have also learned from the study of genetics that the gene pool for any species, though extremely large, is finite or fixed. Therefore, current biology has no plausible explanation for how genetic characteristics not part of the initial gene pool could produce traits of a completely different species or kind. To address this problem, evolutionists have suggested that genetic mutations are the explanation. Unfortunately for the theory, mutations are extremely rare, almost always harmful, and rarely, if ever, beneficial, being limited to the characteristics available in the original gene pool. That's why dog characteristics, or dog genes, are not found in cat gene pools.

Another challenging issue for the theory that mutations could account for how one species evolves into a completely different species or kind is the math. In the DNA replication process, mutations are extremely rare due to the self-correcting mechanisms. Mutations occur, on average, about once in every ten million duplications of a DNA molecule. [138] The odds increase exponentially of getting two consecutive mutations of related genes. The probability of obtaining the necessary number of consecutive mutations to make the kind of changes required to transform one species into a totally different species is beyond comprehension. For example, the probability of achieving just four consecutive mutations is one in ten million billion trillion. But even if you obtained four consecutive mutations, you're nowhere close to getting the number of consecutive mutations needed to make even a small change in a species, much less the kind of change that would be required to transform one species into a different kind. Some scientists who support the theory of evolution acknowledge that the probability factor is a major impediment to their argument.

But hands down, the most difficult question for the evolutionist to answer is, "How did the very first living organism, or first cell, come into being from nonorganic material?" There is simply no credible explanation for how life could spontaneously arise without DNA, RNA, or any of the complex instructions required to build the different kinds of proteins needed for life.

Consider the fact that proteins are composed of 20 different amino acids and that these amino acids are used to build tens of thousands of proteins needed to enable life. Each of the 20 amino acids must be in the correct sequence for the protein to function properly. The individual protein chains, which are found in all living cells, can consist of hundreds of amino acids. How hundreds of amino acids, each with a one in 20 chance of being in the right order, combine in the correct sequence to produce millions of functioning proteins with thousands of different functions is not understood. How thousands of different proteins could simultaneously come together to produce the first life is nowhere close to being understood!

To understand just how improbable—or, more accurately, impossible—it would be to form the tens of thousands of protein chains needed to create a functioning cell, I recommend viewing the YouTube video titled

"Origin: Probability of a Single Protein Forming by Chance." This video calculates the probability and time necessary to assemble a single medium-sized protein chain of 150 residues by random chance. The video is less than 10 minutes long but brilliantly exposes the enormity of the odds and the time needed to produce a single protein molecule from a primordial soup comprised of all the necessary ingredients for life, many of which were likely not available on prehistoric Earth.

After watching this video, try to imagine the probability of getting a simple yeast cell with 6,000 different proteins by chance. A team of researchers from the University of Toronto's Donnelly Centre for Cellular and Biomolecular Research recently discovered, while examining a simple yeast cell that approximately 42,000,000 protein molecules exist in a yeast cell, with 6,000 different protein types encoded by the yeast genome. [139] The average amino acid chain length is approximately 466 residues. [140] Imagine the time required and the probability of producing just one protein with 466 residues. As shown in the above YouTube video, you'd have to increase the probability and time required to produce a 150-residue protein by 20 to the 316th power!

But as big as that number is, it's still smaller than the probability of getting a fully functioning yeast cell by mere chance on a primordial Earth because proteins are just one of numerous other functioning parts in a living cell. The probability of getting a fully functioning yeast cell by chance is going to be a really BIG number!

Another YouTube video that explains the unlikelihood of getting a single protein chain by chance is called "Origin of Life, the Probability of Making a Protein." I highly recommend viewing both videos to understand the enormity of the probability problem, i.e., the probability of getting a single medium-length protein chain by random chance. Now imagine the probability of getting hundreds of protein chains simultaneously, as well as all the other necessary components of a functioning cell, by random chance... it's beyond comprehension.

Since Darwin first proposed his theory, we have learned that the cell is the basic building block of life and that proteins are the workhorses within the cell that perform most of its functions. To appreciate the complexity of the internal functioning of a cell, I recommend watching the YouTube video, "Inner Life of A Cell – Full Version.mkv."

- 29 -

Information Theory and DNA

Information theory was first developed by MIT engineer and mathematician Claude Shannon in the 1940s. According to Scientific American, his MIT master's thesis in electrical engineering has been called the most important of the 20th century. [141] Later, he published a document in 1948 called "The Mathematical Theory of Communications," which has been referred to as the Magna Carta of Information Theory. [142] Shannon's information theory suggests that information and uncertainty are inversely correlated, meaning that as information increases, uncertainty decreases. For example, individual letters contain less information than words, words contain less information than sentences, and sentences contain less information than paragraphs. According to Shannon's theory, the likelihood of obtaining a specific outcome by chance decreases as the complexity of that outcome increases. For example, the probability of randomly generating Darwin's book *On the Origin of Species* is infinitesimally lower than generating just a few words. In fact, we might conclude that it would be impossible to get Darwin's book from a random letter generator, no matter how many times you tried.

Evolutionists suggest that random chance generated something millions of times more complex than Darwin's

Origin of Species. The theory proposes that random chance organized the sequence of 3.2 billion nucleotides in human DNA and sequenced hundreds of billions of long-chain proteins consisting of 20 different amino acids in millions of different species. They also claim that this was all accomplished in a breathtakingly short time, just 4.5 billion years. The YouTube video "Origin: Probability of a Single Protein Forming by Chance" shows how long it would take to make a single medium-length protein chain of only 150 residues…it's far more than 4.5 billion years. [143]

Evolutionists argue that random chance was responsible for correctly sequencing billions of DNA strands in trillions of cells across millions of different organisms. They claim that this process occurred without the need for information, intelligence, or a system for programming the complex machinery we call life. Since the DNA replication system corrects errors, mutations are unlikely to account for the significant changes necessary to transform one species into a completely different kind. In his book, *In the Beginning was Information*, Dr. Werner Gitt, an expert in information systems, draws the following conclusions from the information found in DNA: "Since DNA code has all the essential characteristics of information, there must have been a sender of this information. Since the density and complexity of DNA information is far greater than man's present technology, the sender must be supremely intelligent. Since information cannot originate from matter and is also created by man, man's nature must have a nonmaterial component (spirit)." [144]

The definition of information is "The act of informing, or communicating knowledge or intelligence." [145] Information science tells us that all information comes from intelligent thought. DNA is the most complex information-carrying system known to man. To date, no one has been able to explain how non-intelligent, non-organic matter (dirt, water, and air) could create a single molecule with the capacity to store hundreds of thousands of pages worth of information. DNA has been shown to be the most sophisticated information storage system ever seen, one that far exceeds the world's most sophisticated computers. So great is DNA's storage capacity that it is now being viewed as the best alternative for solving the world's data storage problems. [146]

- 30 -

Evolution
Absence of Rational Answers

We've all heard the question, "Which came first, the chicken or the egg?" This question presents a significant challenge for evolutionists, as they cannot explain how a chicken comes into existence without an egg or how an egg is created without a chicken. The Bible's answer is clear: the chicken came first. God created fully grown chickens and roosters, just as He did with all other animals, insects, fish, birds, and reptiles. They then mated and produced offspring.

The theory of evolution also faces a significant challenge when it comes to the sexes: which came first, male or female? Females can bear offspring, but only with the contribution of a male. How could evolution, operating by random chance, develop male and female genders simultaneously, since a fully functioning female and male would both be required simultaneously to produce offspring? Evolutionary theory offers no satisfactory explanation for how this could happen.

Now consider the staggering probability that random processes not only simultaneously produced male and female genders but did so for every species on Earth—

without any guiding intelligence. According to evolutionary theory, this occurred countless times across millions of species. The likelihood of such a feat happening by chance is beyond comprehension.

It is simply inconceivable that millions of different insects, fish, birds, reptiles, and animals could simultaneously and independently evolve the male and female biology necessary to produce offspring without an intelligent mechanism directing the process. Currently, science is suggesting that the egg preceded the chicken. An article in New Scientist suggests that eggs evolved a billion years ago, but chickens have only been around for 10,000 years. [147] If that is true, what was hatching from these eggs before chickens, and what was making them? For an egg to produce a chicken, we know that all the genetic makeup of a chicken must first reside in the egg, but how does it get there before the first chickens come into existence? How were the eggs incubated since there were no chickens to sit on them?

What is observed in nature is exactly what the Bible says: all species reproduce after their own kind. They are restricted by the information available in their gene pool. This explanation comports with all known evidence. There are no logical or science-based explanations for how the first chicken egg came into existence without the genetic input from a chicken.

Symbiosis refers to any type of interaction or relationship between two different organisms, including mutualism, where both species benefit. One question often raised is how interdependent species, such as those in mutualistic relationships, evolved to rely on one another for survival.

An example of a mutualistic relationship is honeybees and flowers. The honeybee's survival depends on acquiring flower pollen, and flowers depend on insects, like honeybees, for pollination. Neither could survive without the existence of the other.

When examining life on Earth, we observe thousands of mutualistic relationships among plants, insects, birds, fish, mammals, and other living organisms. Many, if not all, could not function without the presence of others. How could these mutualistic relationships have evolved without any intelligent input? Once again, the most logical explanation, and the one that tracks with scientific observation, is the one presented in the Bible: they all resulted from God's creation.

Here's another conundrum for evolutionary theory. How did monarch butterflies evolve with different lifespans? Why does every fourth generation of the monarch butterfly migrate up to 3,000 miles to spawn in the Sierra Madre Mountains in Mexico? Why don't the first, second, or third generations also migrate? Why do the first three generations of these butterflies only live for 4 to 6 weeks, while each fourth generation lives upwards of 6 to 8 months?

How did evolution enable a caterpillar to develop genetic programming for constructing a cocoon and the genetic coding for transforming into a flying insect, completely recreating its original state? How does evolution account for the GPS programming in fourth-generation monarch butterflies, which enables them to locate the spawning grounds in Mexico? Why don't the first three generations of monarch butterflies have this GPS programming?

The answer is that these things didn't happen through evolution. The Bible says in Romans 1:20 (NIV), "For since the creation of the world, God's invisible qualities – his eternal power and divine nature – have been clearly seen, being understood from what has been made, so that men are without excuse."

– 31 –

Intelligent Design Seed Dispersal Systems

What is intelligent design, and how would we recognize it if we saw it? The 5th edition of the American Heritage Dictionary defines intelligent design as "The belief that the physical and biological systems observed in the universe result chiefly from purposeful design by an intelligent being rather than from chance and other undirected natural processes." [148] When we observe man-made objects like a computer, a watch, a car, or a painting, we immediately recognize them as the work of intelligent design. We would never entertain the idea that these items came into existence by random chance. Yet, when we look at our solar system—with its precisely balanced rotational speeds, orbital patterns, planetary masses, and exact distances from one another, all held together by finely tuned gravitational forces—we are told this extraordinary precision was a coincidental outcome from the Big Bang.

Wherever we encounter design, complexity, and information, we instinctively recognize the presence of a designer, even if we never see them. In every aspect of human experience, we assume that created things are the product of intelligence, not random processes. So, is it reasonable to believe that the staggering complexity and sophistication we see in living organisms—which far surpass anything man-made—could be the result of blind chance? Or does such intricacy point to the work of an intelligent Creator?

When we observe the remarkable mechanisms that plants use to disperse their seeds, we must ask: is it more reasonable to attribute these intricate methods to an intelligent designer or to blind chance?

Plants employ at least four primary strategies to ensure their seeds are spread far and wide. One of these methods involves attaching seeds to animals or insects. Seeds from these plants are equipped with tiny hooks that cling to passing animals or insects, eventually falling off and germinating far from the parent plant.

Another method relies on animals ingesting seeds that are so hard they require the softening effect of stomach acid to germinate. What inherent intelligence within the plant could "know" that its seeds needed this process to sprout? Are we to believe that this complex and precise seed dispersal system evolved purely by random chance?

Other plants disperse seeds using spring mechanisms or wind power. For instance, some varieties of pea plants use explosive force to propel their seeds away from the parent plant. Wind-dispersed seeds display some of the most ingenious designs. For example, dandelions employ parachute-like canopies to carry their seeds through the air. The Javan cucumber uses glider-like wings to transport its seeds, while maple trees attach winged structures to their seeds, allowing the wind to carry them up to 100 feet from the parent tree.

The question remains: where did the information, design, and complexity behind these dispersal mechanisms originate? Is it reasonable to conclude that these systems, which display extraordinary ingenuity and even a mastery of aerodynamics, are the product of evolutionary forces and random chance? Or do they point instead to the work of an intelligent designer?

F.C. Payne, in his book, *The Seal of God*, makes 13 statements that he asserts cannot be answered by those who believe nature (evolution) was behind the design of the two-winged seed used by the genus Hakea tree to propagate its species. Here are his statements:

1. If nature is (the) designer, then the tree must have had a desire to propagate its species.

2. The tree must have known its seed was for this purpose and that it would grow.

3. The tree must have known that just to drop its seed would mean they would choke each other.

4. It must have known what few people know that the seed would fall with the same speed as lead.

5. Therefore, it would have to use great wisdom and devise a means of scattering the seed.

6. The means employed shows a knowledge of the wind.

7. The tree must have known that the wind could exert pressure on a flat surface, the wing.

8. The tree, most certainly, would know about the pull of gravity.

9. It was apparently conscious of air pressure and the law of planing.

10. The tree then, combining these laws, concluded that if the wing was set at an angle of 45 (degrees) to the pull of gravity and with a slight tilt, the seed would spin whilst falling.

11. The tree must have known that the spinning action would extend the falling time so that the wind could carry it further.

12. The tree further shows great wisdom in designing its capsule to split open in exactly the correct place to release the two seeds.

13. In the genus Hakea, as with many others, the bulk of the seed is held in the capsule, kept green with sap, and opens only when the sap ceases to flow, after a fire, or when a bough dies. Here again great wisdom is evidenced, for when green it is protected as it is extremely hard and birds cannot destroy it, and when needed to propagate the species, it opens. [149]

These ingenious mechanisms used for seed dispersal all exhibit design, complexity, and intelligence, suggesting they are the product of a mind. Whenever we observe wings, parachutes, gliders, spring mechanisms, or other complex man-made objects, we instinctively attribute them to the work of a designer. How, then, can we conclude, when observing similar items in nature—some far more sophisticated than anything created by humans—that they are the product of random materials coming together by chance?

Mankind has gained remarkable knowledge and inspiration for the design of all kinds of new products by first observing them in nature. Is it reasonable to assume that the designer from whom we are borrowing this ingenuity is random chance and millions of years of natural selection, or might there be another intelligence behind these miraculous designs?

- 32 -

The Multiverse
A Search for Answers

As the scientific community continues to expand its knowledge of the universe, the human biome, the complexity of cells, quantum physics, and atomic structures, it is becoming increasingly difficult to believe that random chance could have produced the precision, design, and complexity we are observing. The idea that what we are seeing could have resulted from random chance is being seriously questioned. The odds against it are astronomical, trillions of times greater than the probability of choosing one marked atom in the entire universe.

Because the odds of this universe coming into existence by random chance keep decreasing with each new scientific discovery, the idea is rapidly losing supporters, even among those with only a basic understanding of mathematical probabilities. Many now believe there isn't enough time, energy, or material in our universe to support the idea that random chance could have produced the level of complexity we see in our universe in such a short time. A solution to address this probability conundrum was needed. The solution arrived in the form of a hypothesis when Hugh Everett, a Princeton Ph.D.

student, suggested that we might be living in one of a multitude of universes. The concept became known as the multiverse. [150] Britannica.com defines the multiverse as, "a hypothetical collection of potentially diverse observable universes, each of which would comprise everything that is experimentally accessible by a connected community of observers." It goes on to say, "...this universe would constitute just a small or even infinitesimal subset of the multiverse." [151]

But is there any evidence that multiple universes exist? According to a March 12, 2023, article titled, "Is the Multiverse Real? The Science Behind 'Everything Everywhere All At Once,'" there is zero evidence for other universes. [152] The biggest misconception about the multiverse is that it's a bona fide theory. Geraint Lewis, Professor of Astrophysics at the University of Sydney, Australia, and author of "Where Did the Universe Come From," said, "It isn't [a theory]—it doesn't really have a mathematical basis—it is a collection of ideas. In the cycle of science, it remains at the hypothesis stage and needs to become a robust proposition before we can truly understand the consequences." [153]

The appeal of the multiverse is that it endeavors to provide an answer to the probability conundrum. Some adherents believe it is the mathematical savior for the "Big Bang" and Darwin's theory of evolution. Since there is not enough time, material, or energy in our universe to account for how all this complexity came about in such a short time, more universes are needed to overcome the probability problem.

You may be wondering how many more universes would be needed to make the probabilities seem plausible. Would a million, a billion, or even a trillion be enough? No, those numbers are nowhere near enough! The only answer that might still lend mathematical credibility to these theories is that there are an infinite number of universes. We just happen to live in the one in infinity of universes where everything needed to support life came together perfectly by random chance.

Even with an infinite number of universes, which evidence doesn't support, it seems extremely unlikely, even impossible, that random chance could ever create our complex universe. When you consider the extreme fine-tuning of gravity, the expansion rate of the universe, and a hundred other laws and constants needed to create and support life, it's beyond reason. But if there were an infinite number of universes in the minds of those who continue to hold to the hypothesis, it just might be possible. Because to their way of thinking, no matter how astronomical the improbability, it looks possible when the denominator is infinity. So, we are left with two choices: either God did it, or there are an infinite number of universes. To date, the overwhelming evidence is on the side of God.

- 33 -

The Origin of Man
When Did We Get Here?

According to current estimates, the Earth is 4.54 billion years old, plus or minus 50 million years. [154] This is based on various dating methods, such as radiometric dating, stratigraphy, carbon dating, and the age of rocks. Worldatlas.com states, "Humans and their ancestors have been inhabiting the Earth for about 6 million years. *Homo sapiens,* who are the modern form of humans, evolved 300,000 years ago from *Homo erectus.* Human civilizations started forming around 6,000 years ago." [155]

If our human ancestors arrived 6 million years ago, it would mean they were here a thousand times longer than human civilizations have existed. We're told that Homo sapiens, believed to be the modern form of humans, have been here for 300,000 years. If that were true, it would imply that they existed for 294,000 years without forming civilizations.

If our human ancestors have roamed the Earth for 6 million years, how did they manage to do it without leaving evidence to prove they were here? There are no mass grave sites, no written records of their history, no cities, no artifacts, no fossils, or any other evidence to substantiate a 6-million-year record of human history. It's hard to believe that they wouldn't have left something. In fact, it's hard to believe that they wouldn't have left thousands, if not millions, of artifacts and other evidence indicating that they inhabited the Earth for millions of years. Try to imagine our ancestors roaming the Earth for six million years and not leaving billions of clues that they were here. Could it be that they weren't here?

What's even more puzzling is that human records appear to date back to the period of Noah's flood, approximately 4,500 years ago. Shouldn't there be more evidence that mankind has been on the planet for 6 million years? The evidence from history, artifacts, ancient ruins, and written records suggests that we've only been here for 4,500 to 6,000 years, which aligns with the timeline in the Bible.

But can't we date the bones of prehistoric man as being hundreds of thousands, if not millions of years old? The simple answer is no. Dating methods, such as carbon 14, have numerous inconsistencies, which make them unreliable. Let's examine some of the more notorious cases where credibility issues emerged.

Nebraska Man: In 1917, Harold Cook discovered a tooth in Nebraska's Upper Snake Creek beds. He shared the information with Dr. Henry Osborn in February 1922. Osborn, along with Dr. William Matthew, attempted to identify the tooth and concluded that it had belonged to an anthropoid ape. [156] A few months later, Science Magazine released an article stating that a man-like ape had been discovered in North America. Was that true? No, it was not. What Cook found was a single tooth. From this tooth, drawings were created based on an artist's interpretation of what this ape-like man and his mate might have looked like. Then, a story detailing possible aspects of the life of this ape-like man was published and released to the public, earning him the name Nebraska Man. It was postulated that he lived a million years ago.

Wikipedia said, "Although Nebraska man was not a deliberate hoax, the original classification proved to be a mistake and was retracted in 1927." [157] What was the mistake that proved to be an error? Well, it turned out that Dr. Osborn and Dr. Matthew had mistaken a pig's tooth for a man's tooth. From that single pig's tooth came an abundance of details regarding the man-like ape that was supposed to be an ancient ancestor of modern man.

Java Man: Pithecanthropus Erectus was an early human fossil discovered in 1891 and 1892 on the island of Java. He was estimated to be between 700,000 and 1,490,000 years old. At the time of the discovery, it was the oldest hominid fossil ever found. [158] But from what was Java Man constructed? He was constructed from a piece of skull, a fragment of a thigh bone, and three molar teeth. They were found scattered among many other bones. With the help of creative artists, Pithecanthropus Erectus was created, and he became a centerpiece in museums around the world. [159, p. 146]

Eugene Dubois was the leader of the excavation team that discovered Java Man. Dubois argued that the fragments discovered were the missing links between apes and humans. According to Wikipedia, these fossils were so controversial that by the end of the 1890s, nearly 80 publications had already been written about them. But despite Dubois' claims, very few accepted Java Man as an actual transitional form between apes and humans. Some dismissed the fossils as apes, while others believed them to be the remains of modern humans. Many scientists believed Java Man to be a primitive side branch of evolution not related to modern humans. [160]

Heidelberg Man: In 1907, a jawbone was discovered approximately 10 miles southeast of Heidelberg, Germany. According to Britannica.com, Homo heidelbergensis was an extinct species of archaic human (genus Homo). His fossils date from 600,000 to 200,000 years ago in Africa, Europe, and possibly Asia. [161] In his book *The Seal of God*, F.C. Payne made this statement regarding Heidelberg Man: "Most scientists rejected it, as of no value. Today exactly similar jaws can be seen in living men anywhere." [162, p. 146]

Neanderthal Man: The remains of this supposed prehistoric man were found in 1856 in Germany, in the Neander Valley. It is believed that Neanderthals split from modern humans somewhere between 315,000 and 800,000 years ago and lived in Eurasia until about 40,000 years ago. According to Wikipedia, "Neanderthal technology was quite sophisticated. It includes the

Mousterian stone-tool industry and the ability to create fire and build cave hearths, make adhesive birch bark tar, craft at least simple clothes similar to blankets and ponchos, weave, go seafaring through the Mediterranean, and make use of medicinal plants, as well as treat severe injuries, store food, and use various cooking techniques such as roasting, boiling, and smoking." [163] F. C. Payne made this statement about Neanderthal Man, "Here again, we have a case of unbelievable deception. From the skull of what is now known to be human, was built one of the most famous of all the ape-men." [164, p. 146] The notable characteristic of this skull is that the forehead runs back in line with the bridge of the nose. Similar skulls can be found anywhere you look in today's population. [164]

Piltdown Man: was described as, "...[a] proposed species of extinct hominin (member of the human lineage) whose fossil remains, discovered in England in 1910-12, were later proved to be fraudulent. Piltdown man, whose fossils were sufficiently convincing to generate a scholarly controversy lasting more than 40 years, was one of the most successful hoaxes in the history of science." [165] This hoax was accepted as true from 1913 until 1953. The supposed 500,000-year-old bones of Piltdown Man were later estimated to be around 50 years old. [166, p. 147]

Why has the world been so accepting and easily deceived by evidence of bone fragments, teeth, or other supposed remains of prehistoric man? Is it because they wanted to believe it? Could we have done the same thing with the age of the Earth? The generally accepted teaching is that our Earth is very old. The current estimate is 4.54 billion years. But is there any scientific evidence that the Earth is actually young, possibly very young?

- 34 -

The Earth's Age
Surprising Evidence

Does it matter whether the Earth is old or young? Yes, it does. The answer is important because the credibility of the Bible is tied to a young Earth. The genealogy from Adam to Christ only accounts for about 4,000 years of human history. If evolution were true, then God's Word, which states that He created the world and all that is in it in six literal days, would be untrue; and if that is untrue, the whole text becomes suspect.

Modern science believes that all life on Earth evolved from inorganic matter over millions of years. But modern science is unlocking new information on the structure of the universe, the complexity of cells, the mysteries of quantum physics, proof of intelligent design in nature, and compelling evidence that the Earth is thousands, not millions of years old. Evidence against evolution and the Big Bang theory is rapidly mounting and beginning to validate the timeline of the Bible.

Significant and compelling evidence is accumulating that suggests the Earth is young, perhaps very young. According to Dr. Russell Humphreys, who earned a Ph.D. in Physics from LSU, there are over a hundred different dating methods used to determine the age of the Earth, and about 90 percent of these indicate that the Earth is young. According to Humphreys, most dating methods suggest the Earth is young, not billions of years old. [167]

What follows is a short list of the evidence that supports the biblical view that the Earth is young, even very young.

The degeneration of the human genome points to a young Earth. Recent human genome research has shown that human beings are devolving, not evolving. In other words, the human genome is losing information and accumulating more mutations over time. Contrary to evolutionary theory, we are not evolving into higher-level beings over time. The generations that followed Adam and Eve experienced progressive degeneration or decay in their DNA, resulting in a loss of genetic information. Dr. Tomkins, in his article "Six Biological Evidences for a Young Earth," says, "...empirical genetic clocks determined by both secular and creation researchers indicate a beginning point of human variation associated with degeneration starting about 5,000 to 10,000 years ago." [168]

The age of Mitochondria Eve suggests a very young Earth. Inside human cells are small organelles called mitochondria. These are the energy factories of the cell. They produce a substance known as ATP (Adenosine Triphosphate). "Scientists have studied mitochondria DNA in people groups around the world and discovered the data are consistent with a single origin of all humans less than 10,000 years ago." [169]

In addition to the young age of mitochondria, genetic researchers have also discovered a limited amount of variation in the DNA sequence of the human Y chromosome across the world population. This limited variation in the Y chromosome is consistent with the origin of humanity dating to 4,000 BC. The genetic research on Y chromosomes in humans aligns with the biblical age of the Earth. In their research paper titled "An Overview of the Independent Histories of the Human Y Chromosome and the Human Mitochondrial Chromosome," Robert W. Carter, Stephen S. Lee, and John C. Sanford made this comment: "The genetic evidence strongly suggests that Y Chromosome Adam/Noah and Mitochondrial Eve were not just real people, they were the progenitors of us all. In this light, there is every reason to believe that they were the Adam/Noah and Eve of the Bible." [170] In this same article, the authors drew the following conclusion, "Strict Darwinists and theistic evolutionists both claim the biblical Adam and Eve never existed. However, after carefully considering the information provided here, their case is significantly weakened. In fact, if the Bible were not true, one would never expect such a strong concordance between biblical and phylogenetic history, as we have shown." [171]

The discovery of soft tissue, blood vessels, blood cells, bone cells, and DNA in dinosaur fossils, like those found by Mary Schweitzer in a Tyrannosaurus rex specimen, provides additional reasons to believe in a young Earth. Brian Thomas, a scientist from the Institute for Creation Research (ICR), compiled a list of 41 journal papers that describe soft tissues and biomolecules found in the fossils of various land and sea animals and plants. [172] "Many of these findings were made and documented by secular scientists. Some of these discoveries involve fossils alleged to be 250 to more than 550-plus million years old. Because it would be impossible for these highly degradable compounds to last for more than a few thousand years, the evidence clearly points to a young age for Earth and to the global flood that produced the fossilized remains, burying them quickly in sediments about 4,500 years ago." [173]

Another reason to believe in a young Earth is the abundance of living fossils, which inexplicably have remained unchanged. Dr. Tomkins at the Institute of Creation Science states, "One living fossil tree, the Wollemi pine, supposedly first showed up in the fossil record over 200 million years ago and not only still exists but has living specimens dated at less than 1,000 years. The lack of evolution observed in living fossils, combined with their sudden appearance in the fossil record and then absence for millions of years, doesn't support the evolutionary paradigm. Instead, the fossil record shows that a global flood occurred only thousands of years ago and progressively buried ecosystems." [174] He goes on to state, "The geological column and its accompanying fossils represent the progressive extinctions and rapid ebb and flow of water burying entire ecosystems over the course of the flood year." [175]

The current world population provides further evidence pointing to a young Earth and aligns with the biblical account of man's time on Earth. In 2015, Robert Carter and Chris Hardy did computer modeling for the Earth's population growth that included multiple variables, such as the age of reaching maturity, minimum child spacing between births, and the age of menopause. They also factored in probabilities like polygamy, twinning rates, and the risk of death according to age. Their conclusion was that "it is trivial [i.e., no great difficulty] to obtain the current world population from three founding couples in four and a half millennia." [176] Their verdict aligns with the repopulation of the Earth, beginning with Noah, his sons, and their wives. If human beings have existed for six million years, as evolutionists believe, why isn't the current population much higher? And why haven't we discovered millions of graveyards, billions of skeletons, or evidence of these human ancestors in the fossil record?

New DNA discoveries point to a young Earth. DNA can only survive at most 10,000 years, and then only if maintained in pristine conditions. [177] According to an ICR article titled 'DNA in Dinosaur Bones,' "The moa research team measured the half-life of DNA to be 521 years under average local temperatures. After this time, only half of the DNA present when the animal died should remain. And after another 521 years, only half of that remains, and so on until none is left. At this rate, DNA molecules in bone break down after only 10,000 years into tiny chemical segments too short for modern technology to sequence. And this result assumes preservation factors that optimize biochemical longevity." [178] If DNA has been found in dinosaur bones, it's further evidence that the Earth is young.

Carbon-14 radioactive dating suggests that the Earth is much younger than 4.54 billion years. The half-life of carbon-14 is 5,730 years. [179] Therefore, we would not expect to find carbon-14 in anything dated more than a few hundred thousand years. Yet carbon-14 has shown up in items supposedly hundreds of millions of years old, like diamonds. [180] The presence of carbon-14 in diamonds suggests the age of the Earth is young.

Comets are large objects composed of dust and ice and are believed to be remnants of the formation of our solar system about 4.6 billion years ago. [181] According to Jason Lisle, Ph.D., an astrophysicist from the University of Colorado, "...a typical comet can last no more than 100,000 years. [182] If that is true, then the age of our solar system and the age of the Earth may be significantly younger than current estimates.

The presence of intergalactic dust on the Moon's surface suggests a young Earth, Moon, and solar system. Scientists are able to measure the rate at which dust accumulates on the Moon's surface. Based on these estimates, there was concern that America's first Moon landing might be jeopardized by the amount of dust that would have accumulated on the moon's surface over billions of years. Andrew Snelling, Ph.D., and David Rosh, M.S., in their article "Moon Dust and the Age of the Solar System," state,

"Prior to direct investigations, there was much debate amongst scientists about the thickness of dust on the moon. Some speculated that there would be very thick dust into which astronauts and their spacecraft might 'disappear'..." They went on to say, "One of the evidences for a young earth that creationists have been using now for more than two decades is the argument about the influx of meteoritic material from space and the so-called "dust on the moon" problem. They pointed out that the best measurements of the influx of meteoritic material from space have been made by Hans Pettersson, and based on his calculations, there would be 182 feet of this dust on the surface of the Earth if it were 5 billion years old. [183] The lack of intergalactic dust on the surfaces of the Earth and Moon suggests that the Earth is much younger than previously thought.

The half-life of the Earth's magnetic field provides compelling evidence for a young Earth. Measurements of the Earth's magnetic field show that it is losing 50% of its energy every 1,400 years. John D. Morris, in his article "Earth's Magnetic Field," stated, "Calculating back into the past, the present measurements indicate that 1,400 years ago, the field was twice as strong. It continues doubling every 1,400 years, until about 10,000 years ago it would have been so strong the planet would have disintegrated—its metallic core would have separated from its mantle. The inescapable conclusion we can draw is that the Earth must be fewer than 10,000 years old." [184]

Dr. Humphreys, in his article, "The Creation of Cosmic Magnetic Fields," postulates, "After 6,000 years of decay, including energy losses from magnetic reversals during the Genesis Flood, the strength of the Earth's magnetic field would be what we observe today." In the conclusion of his article, he said, "Magnetic fields, or vestiges of them, appear to exist everywhere throughout the cosmos. The water-origin theory offers an explanation which works quantitatively over a very wide range of phenomena. For most of the bodies in the solar system, the theory only works for an age of about 6,000 years. Throughout the cosmos, it only works for a water origin, not for the other materials that now constitute most of the heavenly bodies, such as hydrogen, silicon, iron, and so forth. The agreement of theory and observations thus strongly supports the biblical account of creation. God may have left us magnetic fields in the heavens as evidence of His handiwork." [185]

Mercury's magnetic field suggests a much younger solar system than previously theorized. The U.S. satellite Mariner 10 measured Mercury's magnetic field in 1975, and the Messenger spacecraft took another reading in 2008. The measurements showed that there had been significant degradation in the energy of Mercury's magnetic field over a period of just 33 years. [186] If the magnetic fields of the planets in our solar system have half-lives like that of the Earth, how is it possible that strong magnetic fields still exist if our solar system is billions of years old? These facts argue for both a young Earth and a young solar system.

The concentration of helium in the Earth's atmosphere argues for a very young Earth. Dr. Melvin A. Cook was one of the first to point out the disparity between the Earth's age and the atmospheric helium content in his 1957 article, "Where is the Earth's Radiogenic Helium?" The release, or diffusion rate, of helium into the Earth's atmosphere can be measured, and the amount being released exceeds that escaping from our atmosphere. "Dr. Larry Vardiman, an ICR atmospheric scientist, has shown that even after accounting for the slow leakage into space, the Earth's atmosphere has only about 0.04% of the helium it should have if the Earth were billions of years old." [187]

Because helium diffuses very rapidly, all the world's helium should have been released into the atmosphere in less than 100,000 years. [188] "Even though some helium escapes into outer space, the amount still present is not nearly enough if the Earth is over 4.54 billion years old. In fact, if we assume no helium was in the original atmosphere, all the helium would have accumulated in only 1.8 million years, even from an evolutionary standpoint. But when the catastrophic Flood upheaval is factored in, which rapidly released huge amounts of helium into the atmosphere, it could have accumulated in only 6,000 years." [189]

The above evidence for a young Earth coincides with the biblical account of the age of the Earth at approximately 6,000 years. Recent scientific discoveries confirm and validate the truth of God's Word.

- 35 -

Man's Science Versus God's Science

Man's understanding of science is always evolving, but the principles and laws that govern science remain constant because they were established by God. The God who created science does not make mistakes. His timeless word, which is thousands of years old, remains unchanged and is as accurate today as it was when it was first inspired. In contrast, scientific knowledge is constantly updated as new facts are discovered. Textbooks and scientific theories are revised to reflect the latest new discoveries, leading to a process of continual change. What was once accepted as true may later be proven false, and vice versa.

In Aristotle's day, scientists believed that the four main components of the known universe were earth, water, air, and fire. It took science nearly six thousand years and a lot of new scientific discoveries to conclude that the universe is best described by time, space, matter, and energy. Had we paid closer attention, we could have come to that conclusion six thousand years earlier by simply reading the first sentence in the Bible, Genesis 1:1 (NIV), "In the beginning (time), God created (energy) the heavens (space) and Earth (matter)." The Bible states that God, not a Big Bang, was the creator of our universe. We read in Isaiah 45:18 (KJV), "For thus saith the LORD that created the heavens; God himself that formed the Earth and made it; he hath established it; he created it not in vain; he formed it to be inhabited..."

Up until the 5th century, our forefathers believed that the Earth was flat, that it was the center of the universe, that the sun orbited the Earth, and many believed that it rested on the back of a very large turtle. These ideas were accepted as truth, even though God's Word had told us three thousand years earlier that the Earth is spherical and hangs on nothing.

In the Book of Job 26:7 (NIV), we read, "He spreads out the northern skies over empty space; he suspends the earth over nothing." The Bible also says in Isaiah 40:21-22 (NIV), "Do you not know? Have you not heard? Has it not been told you from the beginning? Have you not understood since the earth was founded? He sits enthroned above the circle of the earth...He stretches out the heavens like a canopy, and spreads them out like a tent to live in." This is also the first mention of an expanding universe that science only recently discovered.

The Bible also informed us that life is in the blood, thousands of years before science discovered this. In Leviticus 17:11 (NIV), we read, "For the life of a creature is in the blood, and I have given it to you to make atonement for yourselves on the altar; it is the blood that makes atonement for one's life." It wasn't until the late 1800s that science began to understand the life-giving importance of blood. "At the end of the 1800s, laboratory science introduced a new (microscopic) view of blood that led to our modern scientific understanding of the fluid and its function in the body." [190]

Scientists learned in 1991 that the heart has neurons like the brain. Who would have thought that? But God knew! The Bible spoke of the thoughts of the heart thousands of years earlier. In Genesis 6:5 (NIV), we read, "The LORD saw how great man's wickedness on the earth had become, and that every inclination of *the thoughts of his heart* was only evil all the time."

Dr. Andrew Armour, a neurologist, was the first to discover neurons in the heart, which he referred to as the "little brain in the heart." His discoveries were made in the early 1990s. After his discovery, it was found that the neurons in the heart have both long-term and short-term memory. After receiving a heart transplant, some patients have reported experiencing memories and personality traits that appear to have been inherited from their donors. Solomon asks God for wisdom in the Book of 1 Kings, and in 1 Kings 3:12 (NIV), it is written, "I will do what you have asked. I will give you a *wise and discerning heart.*"

When we think of wisdom and discernment, most of us assume that they are attributes of the mind or brain. Not until Dr. Armour's discovery of the neurons in the heart did the Scriptures that reference the thoughts of the heart make sense from a scientific perspective. In 1 Kings 10:24 (NIV), the Bible says, "The whole world sought audience with Solomon to hear the wisdom God had put in his heart." In 1 Chronicles 29:18 (KJV), we read, " O LORD God of Abraham, Isaac, and of Israel, our fathers, keep this forever in the imagination of the *thoughts of the heart* of thy people, and prepare their heart unto thee:"

Instruction is something that we have assumed is only a function of our brain, but the Bible says in Psalm 16:7 (NIV), "I will praise the Lord, who counsels me; even at night my heart instructs me." Psalm 19:14 speaks of the meditation of the heart, and in Psalm 40:8 (NIV), referring to King David, it says, "...your law is within my heart."

Additionally, in the Book of Psalms 44:21 (NIV), we find these words, "would not God have discovered it since he knows the secrets of the heart?" Haven't we always assumed that any secrets we held were stored away in our brains?

Has science not assumed from the foundation of the Earth that understanding is a function of the brain? Has any scientist in the past 6,000 years, prior to the discovery of neurons in the heart, suggested that understanding could also be a function of the heart? But God knew this and inspired King David to write these words in Psalm 49:3 (NIV), "My mouth will speak words of wisdom; the utterance from my heart will give understanding."

Here are six more Bible verses that reference the thoughts of the heart.

1 Chronicles 29:18, (NIV) "Lord, the God of our fathers Abraham, Isaac and Israel, keep these desires and thoughts in the hearts of your people forever, and keep their hearts loyal to you."

Matthew 15:19 (NIV), "For out of the heart come evil thoughts—murder, adultery, sexual immorality, theft, false testimony, slander."

Mark 7:21 (NIV), "For from within, out of men's hearts, come evil thoughts, sexual immorality, theft, murder, adultery, greed, malice, deceit, lewdness, envy, slander, arrogance, and folly.

Romans 2:15 (NIV), "...they show that the requirements of the law are written on their hearts, their consciences also bearing witness, and their thoughts sometimes accusing them and at other times even defending them."

Hebrews 4:12 (NIV), "For the word of God is living and active. Sharper than any double-edged sword, it penetrates even to dividing soul and spirit, joints and marrow; it judges the thoughts and attitudes of the heart."

Here is an area where science was sure the Bible was incorrect. Leviticus 11:6 (NIV) states, "The rabbit, though it chews the cud, does not have a divided hoof; it is unclean for you." Many still believe this is one of the best examples of where the Bible is in error. However, we now know that the rabbit produces two types of droppings: one of which is cecotropes, partially digested food, and the other is waste. When the rabbit eats or chews the cecotropes, it is essentially completing the digestion process, just like other animals that chew the cud complete their digestive process for partially digested food. When we examine the original Hebrew text, we find that the word that was translated as "cud" is the Hebrew word gêrâh, which means something that has been swallowed. [191] Rather than chewing the cud, the translation from Hebrew to English should state that the rabbit chews something previously swallowed.

One of the main areas where science and the Bible are most at odds is the question of how humanity came into existence. According to the Bible, man was created by God. It says in Genesis 1:26-27 (NIV), "Then God said, 'Let us make man in our image, in our likeness' ... So God created man in his own image, in the image of God he created him; male and female he created them." However, the prevailing view today is that all life, from the smallest bacteria to Homo sapiens, self-generated from inorganic matter through a process known as abiogenesis. This idea contradicts the entire history of observed science since no one has ever witnessed life emerging from non-life. The basis for this theory is extremely weak and becoming weaker with every new scientific discovery.

The brightest minds on Earth have tried and failed to create or replicate the very simplest life forms. However, no one has come close. This is because there are no simple life forms. The simplest known bacterium, Mycoplasma genitalium, has 485 protein-coding genes, and the whole synthetic genome contains over 500,000 base pairs. [192]

A team of bioengineers led by Stanford's Markus Covert, succeeded in modeling this bacterium. "What is fascinating is how much horsepower they needed to partially simulate this simple organism. It took a cluster of 128 computers running for 9 to 10 hours to generate the data on the 25 categories of molecules that are involved in the cell's lifecycle processes." [193]

With each passing year, science adds volumes of new complexities to the question of how life might have spontaneously generated from inorganic materials. The more we learn about the staggering complexity of life—including forms we once considered simple—it becomes increasingly challenging to explain how abiogenesis could have produced the first living organism. It's equally difficult to imagine how evolutionary processes and random chance could have transformed that original organism into the complex beings we are today. Rather than bringing us closer to answers, new scientific discoveries only deepen the complexity, making such explanations much harder to accept.

James Tour, a synthetic chemist and world-renowned expert in nanotechnology, points out that four classes of macromolecules are needed for life: carbohydrates, proteins, nucleic acids (like RNA and DNA), and lipids. While some building blocks of these molecules have been synthesized in laboratory conditions and detected in nature, there is no conclusive evidence that all four existed together in functional forms on a prebiotic Earth. Furthermore, no known naturalistic mechanism explains the origin of the complex information systems or instructional processes necessary for the first living cells to emerge.

The current scientific view suggests that complex life forms, emerged from purely natural processes acting on inorganic matter—without any guiding intelligence, purpose, or pre-existing instructions. However, when we analyze the conditions required for life, the informational complexity within DNA, and the fine-tuning of the universe, the probability of such an outcome occurring by chance appears astronomically low. In contrast, the hypothesis of an intelligent Creator provides a coherent explanation for the origin of life and the intricate order we observe, even in the simplest life forms. Given these considerations, one could argue that belief in unguided processes requires an even greater leap of faith than belief in God.

The Bible states that God created all life forms and that they reproduce after their own kind, which is exactly what we witness everywhere. There are no examples of life evolving or transitioning from one kind of animal to another kind. However, there are hundreds of thousands of examples of micro-evolution, where small changes occur within a kind or within a species. For example, dogs, wolves, coyotes, foxes, hyenas, etc., are all from the same family or kind and could produce new varieties within the dog family through breeding. But since the gene pool is limited to dog characteristics, you won't see cat or horse characteristics or anything but dog features showing up within the dog gene pool.

If evolution were a fact, there would be millions of transitional links between kinds or between species within the fossil record. That's exactly what Charles Darwin believed would someday be found. In fact, Darwin said that a possible objection to his own theory was the lack of transitional forms in the fossil record: "Why, if species have descended from other species by fine gradations, do we not everywhere see innumerable transitional forms?" [194] Could it be that it didn't happen that way?

Given the evidence in Scripture and science for the accuracy and reliability of God's word, what are the implications for mankind?

- 36 -

What the Bible Says About Sin

According to the Bible, God hates sin, and anyone who is a follower of Christ will hate it too. In Romans 6:1-4 (KJV), the apostle Paul proclaims, "What shall we say then? Shall we continue in sin, that grace may abound? God forbid. How shall we, that are dead to sin, live any longer therein? Know ye not, that so many of us as were baptized into Jesus Christ were baptized into his death? Therefore, we are buried with him by baptism into death: that like as Christ was raised up from the dead by the glory of the Father, even so we also should walk in newness of life." He goes on to say in Romans 6:6-7 (KJV), "Knowing this, that our old man is crucified with him, that the body of sin might be destroyed, that henceforth we should not serve sin. For he that is dead is freed from sin." Romans 6:11-12 (KJV) says, "Likewise reckon ye also yourselves to be dead indeed unto sin, but alive unto God through Jesus Christ our Lord. Let not sin therefore reign in your mortal body, that ye should obey it in the lusts thereof." Romans 6:14 (KJV) says, "For sin shall not have dominion over you: for ye are not under the law, but under grace." And Romans 6:18 (KJV) says, "Being then made free from sin, ye became the servants of righteousness." The idea conveyed in the above Scriptures is that Christians are to die to sin and become servants of righteousness.

Professing Christians who find themselves in habitual sin should consider carefully the following Scriptures.

1 John 3:6 (NIV) says, "No one who lives in him keeps on sinning. No one who continues to sin has either seen him or known him."

1 John 3:9 (NIV) says, "No one who is born of God will continue to sin, because God's seed remains in him; he cannot go on sinning, because he has been born of God.

1 Peter 2:24 (NIV) says, "'He himself bore our sins in his body on the tree, so that we might die to sins and live for righteousness…"

Hebrews 10:26-29 (NIV) says, "If we deliberately keep on sinning after we have received the knowledge of the truth, no sacrifice for sins is left, but only a fearful expectation of judgment and of raging fire that will consume the enemies of God. Anyone who rejected the law of Moses died without mercy on the testimony of two or three witnesses. How much more severely do you think someone deserves to be punished who has trampled the Son of God underfoot, who has treated as an unholy thing the blood of the covenant that sanctified him, and who has insulted the Spirit of grace?"

Hebrews 6:4-6 (NIV) says, "It is impossible for those who have once been enlightened, who have tasted the heavenly gift, who have shared in the Holy Spirit, who have tasted the goodness of the word of God and the powers of the coming age, if they fall away, to be brought back to repentance, because to their loss they are crucifying the Son of God all over again and subjecting him to public disgrace."

Galatians 5:16 (NIV) says, "...live by the Spirit and you will not gratify the desires of the sinful nature."

Many preachers, pastors, priests, and others have suggested that we are all sinners saved by grace, and this is a true statement. However, where I think some churches may be missing the mark is when they use the passages in Romans 7:14-25 to suggest that if we find ourselves in habitual sin, we are no different than Paul, who said in Romans 7:14-15 (NIV), "...but I am unspiritual, sold as a slave to sin. I do not understand what I do. For what I want to do I do not do, but what I hate I do." He then goes on to say in verses 18 and 19 of Romans 7 (NIV), "...For I have the desire to do what is good, but I cannot carry it out. For what I do is not the good I want to do; no, the evil I do not want to do – this I keep on doing." He concludes the chapter by saying in Romans 7:24 (NIV), "What a wretched man I am! Who will rescue me from this body of death? Thanks be to God – through Jesus Christ our Lord!"

It is hard to reconcile how the same man who wrote the above words in Romans 7 was also the author of Romans 6:2b (NIV), where he said, "...We died to sin: how can we live in it any longer?" How then does one explain that the same man who wrote in Romans 7:19b (NIV), "...the evil I do not want to do – this I keep on doing," also wrote

Romans 6:6-7 (NIV), "For we know that our old self was crucified with him so that the body of sin might be rendered powerless, that we should no longer be slaves to sin – because anyone who has died has been freed from sin." In verse 6:18 (NIV) Paul says, "you have been set free from sin and have become slaves to God." He concludes with Romans 6:22-23 (NIV) saying, "But now that you have been set free from sin and have become slaves to God, the benefit you reap leads to holiness, and the result is eternal life."

I believe the key to unlocking the truth in these seemingly contradictory words in Romans chapter 7 can be found in Romans 6:14 (NIV), which states, "For sin shall not be your master, because you are not under law, but under grace." For it is those who are under the law that have no power over sin. We read in 1 Corinthians 15:56 (NIV), "The sting of death is sin, and the power of sin is the law." The Jews have been living under the Law of Moses (The Ten Commandments and over 600 additional laws) for more than 3,400 years. But the law did not give them power over sin; it simply showed them where they were going wrong and where they were breaking the law. But when Christ died as the final sacrifice for sin, he broke the power of the law for those who put their faith in Him and submit to His Lordship.

Romans chapter 7 (NIV) opens with these words, "Do you not know brothers – for I am speaking to men who know the law – that the law has authority over a man only as long as he lives?" He then goes into an explanation of how the law works, using marriage as an example. Finally, he states in Romans 7:4 (NIV), "So, my brothers, you also died to the law through the body of Christ..." He goes on to say in verses 5 and 6, "For when we were controlled by the sinful nature, the sinful passions aroused by the law were at work in our bodies, so that we bore fruit for death. But now, by dying to what once bound us, we have been released from the law so that we serve in the new way of the spirit, and not in the old way of the written code (the law)."

After making the above statements, Paul explains what it was like to live under the law. In Romans 7:8-10, he says, "But sin, seizing the opportunity afforded by the commandment (the law), produced in me every kind of covetous desire. For apart from the law, sin is dead. Once I was alive apart from the law (before he knew the commandments); but when the commandment came (when he became aware of them), sin sprang to life and I died. I found that the very commandment that was intended to bring life actually brought death." In Romans 7:19 (NIV), Paul is explaining his life under the law. When he said, "For what I do is not the good I want to do; no, the evil I do not want to do – this I keep on doing." Paul was not describing his new nature in Christ, because the law lacks the ability to transform one's soul and spirit. Only Christ can bring new life and, with it, the power to overcome sin. Paul concludes by saying, "Who will rescue me from this body of death? Thanks be to God – through Jesus Christ our Lord!" (Romans 7:24 NIV)

Paul then returns to his original premise from Romans chapter 6, where he said in verse 14 (NIV), "sin shall not be your master, because you are not under law, but under grace." Then Paul begins chapter 8 by explaining the difference between the Law of Sin and Death (The Law of Moses) and the Law of the Spirit of Life (The New Life in Christ). He goes on to say in Romans 8:3-4 (NIV), "For what the law was powerless to do, in that it was weakened by the sinful nature, God did by sending his own Son in the likeness of sinful man to be a sin offering. And so, he condemned sin in sinful man, in order that the righteous requirements of the law might be fully met in us, who do not live according to the sinful nature but according to the Spirit." In verse 7, we read, "You, however, are controlled not by the sinful nature but by the Spirit, if the Spirit of God lives in you."

Paul continues this theme in Romans 8:12-13 (NIV), "Therefore, brothers, we have an obligation – but it is not to the sinful nature, to live according to it. For if you live according to the sinful nature, you will die, but if by the Spirit you put to death the misdeeds of the body, you will live." This idea is reinforced in Galatians 5:24 (NIV) where it says, "Those who belong to Christ Jesus have crucified the sinful nature with its passions and desires."

My concern is that some churches may be misleading their congregations by suggesting that when Christians sin, they are no different from Paul the Apostle. They use Paul's words in Romans 7 to support the idea that Christians who find themselves in habitual sin are just like Paul. Romans chapter 7 is a description of Paul's life before he was saved, when he was under the Law of Sin and Death. After salvation, he was under the Law of the Spirit of Life.

We must not compromise when it comes to sin. That is not the nature of true Christians. We are to be a holy people, set apart from this world. There must be a clear distinction between the children of God and the children of Satan. Paul the sinner, portrayed in Romans chapter 7, was the same Paul who held the coats of those who stoned Stephen. He was the same Paul who habitually persecuted Christians. Romans chapter 7 describes Paul's life under the law before he was transformed by the Spirit of God. The Paul who wrote chapters 6 and 8 in Romans is the same Paul who wrote in II Corinthians 7:1 (NIV), "Since we have these promises, dear friends, let us purify ourselves from everything that contaminates body and spirit, perfecting holiness out of reverence for God."

Why is holiness so important to the Christian way of life? One reason is because Hebrews 12:14 says, "Without holiness no one will see God." I believe another reason is because it enables God's power to work through us. The prophets of the Bible were holy men. The disciples of Jesus became holy men, and God's power miraculously worked through them to heal the sick, raise the dead, and cast out demons. I Corinthians 4:20 (NIV) says, "For the Kingdom of God is not a matter of talk, but of power." And we read in II Timothy 1:7 (NIV), "For God hath not given us the spirit of fear; but of power..." Notice that the Holy Spirit, when inspiring these verses, emphasized power.

The church and the followers of Christ were clearly meant to be endowed with power. How can we expect the power of God to flow through vessels of sin? If we want to see the power of God manifested, we must sell out to Him and fully submit to His will. He must become the Lord of everything in our lives. For it is in holy vessels that God's power flows. If we do not see the power of God manifested in today's church, it may be because we have embraced sin as if it were a natural part of Christian life. If we are relating to the Paul who said in Romans 7:19 (NIV): "For what I do is not the good I want to do; no, the evil I do not want to do – this I keep on doing" we are misinterpreting his message.

- 37 -

Fear of the Lord
The Foundation of Wisdom

In the Book of Job, chapter 28, there are 17 verses that discuss wisdom and understanding. In these verses, Job compares wisdom and understanding to gold, silver, crystal, rubies, and other precious items. We learn from Job that wisdom and understanding are more valuable than all of these in God's eyes. Referring to wisdom and understanding, Job 28:13 (NIV) says, "Man does not comprehend its worth." And in Job 28:17 (NIV), we read, "Neither gold nor crystal can compare with it." Regarding wisdom and understanding, Job 28:23 (NIV) says, "God understands the way to it and he alone knows where it dwells." The chapter concludes with this statement in Job 28:27-28 (NIV), "Then he (God) looked at wisdom and appraised it; he confirmed it and tested it. And he said to man, The fear of the Lord – that is wisdom, and to shun evil is understanding."

The Bible associates the fear of the Lord with wisdom and shunning evil with understanding. There are many Scriptures that describe wisdom and understanding and explain their benefits. There is also a wealth of Scriptures that describe what one must do to acquire them. The Scriptures below from the Book of Proverbs associate the fear of the Lord with wisdom and understanding and describe some of their benefits.

Proverbs 1:7 (NIV), "The fear of the Lord is the beginning of knowledge, but fools despise wisdom and discipline."

Proverbs 8:13 (NIV), "To fear the Lord is to hate evil."

Proverbs 9:10 (NIV), "The fear of the Lord is the beginning of wisdom, and knowledge of the Holy One is understanding."

Proverbs 10:27 (NIV), "The fear of the Lord adds length to life, but the years of the wicked are cut short."

Proverbs 14:27 (NIV), "The fear of the Lord is a fountain of life, turning a man from the snares of death."

Proverbs 14:16 (NIV), "A wise man fears the Lord and shuns evil."

Proverbs 15:33 (NIV), "The fear of the Lord teaches a man wisdom, and humility comes before honor."

Proverbs 19:23 (NIV), "The fear of the Lord leads to life; then one rests content, untouched by trouble."

Proverbs 22:4 (NIV), "Humility and the fear of the Lord bring wealth and honor and life."

Instructions for obtaining wisdom, understanding, and the knowledge of God can be found in Proverbs 2:1-4 (NIV), which says, "My son, if you accept my words and store up my commands within you, turning your ear to wisdom and applying your heart to understanding, and if you call out for insight and cry aloud for understanding, and if you look for it as for silver and search for it as for hidden treasure, then you will understand the fear of the Lord and find the knowledge of God."

The phrases "fear God," "fear the Lord," and "the fear of the Lord" appear hundreds of times in the Bible. [195] Is it possible that the absence of "the fear of God" in today's society is one of the primary reasons the world has disintegrated into its present condition? Do churches today teach their congregations why they should fear God, or is all the focus on God's grace, love, mercy, and forgiveness? Clearly, grace, love, mercy, and forgiveness are key attributes of God's character, but are we ignoring why we should also teach the fear of the Lord?

The Bible says in Matthew 7:13 (NIV), "Enter through the narrow gate. For wide is the gate and broad is the road that leads to destruction, and many enter through it. But small is the gate and narrow the road that leads to life, and only a few find it." And we read in Matthew 7:21-23 (NIV), "Not everyone who says to me, 'Lord, Lord,' will enter the kingdom of heaven, but only he who does the will of my Father who is in heaven. Many will say to me on that day, 'Lord, Lord, did we not prophesy in your name, and in your name drive out demons and perform many miracles?'

Then I will tell them plainly, 'I never knew you. Away from me, you evildoers.'" Consider these words in Psalm 89:7 (NIV), "In the council of the holy ones God is greatly feared..." The previous verses clearly tie the fear of the Lord to wisdom and understanding.

Has anyone contemplated who comprises the council of the holy ones? Who are its members? Surely, they must be among the elite of the elite, the most highly esteemed and righteous of God's elect. Why would the Holy Spirit inspire these words, "In the council of the holy ones God is greatly feared..." Why did Jesus delight in the fear of the Lord? We know He did because the following words are recorded in Isaiah 11:1-3 (NIV), "A shoot will come up from the stump of Jesse; from his roots a Branch will bear fruit. The Spirit of the Lord will rest on him – the Spirit of wisdom and of understanding, the Spirit of counsel and of power, the Spirit of knowledge and of the fear of the Lord – and he will delight in the fear of the Lord." Note that the above Scriptures from Isaiah identify seven spirits that will rest on Jesus and that among them are wisdom and understanding and the fear of the Lord.

Psalm 111:10 (NIV) says, "The fear of the Lord is the beginning of wisdom." Proverbs 16:6 (NIV) says, "Through love and faithfulness sin is atoned for; through the fear of the LORD a man avoids evil." The Scriptures referred to earlier in the Book of Proverbs describe many of the benefits associated with the fear of the Lord; but how does it help one keep from sinning? The key is to understand that there are consequences associated with our sins.

Since God is just, righteous, and holy, it is in His nature to judge mankind for our actions, deeds, and even our thoughts. He also gave us commandments to live by in both the Old and New Testaments. In Matthew 5:27-30 (NIV) Jesus said, "You have heard that it was said, do not commit adultery. But I tell you that anyone who looks at a woman lustfully has already committed adultery with her in his heart! If your right eye causes you to sin, gouge it out and throw it away. It is better for you to lose one part of your body than for your whole body to be thrown into hell."

From Genesis to the Book of Revelation, the Bible tells the story of man's fall and God's plan for redemption. In Genesis chapter 2, we read the account of man's first sin: breaking God's command to not eat from the tree of the knowledge of good and evil. The extreme consequences for Adam and Eve, and all of mankind for breaking this commandment, are incomprehensible. Their sin affected all of human history and brought untold sorrow and suffering into a world that had previously been sinless and perfect in every way.

In Genesis 6:5-8 we read, "The LORD saw how great man's wickedness on the earth had become, and that every inclination of the thoughts of his heart was only evil all the time. The LORD was grieved that He had made man on the earth, and His heart was filled with pain. So the LORD said, 'I will wipe mankind, whom I have created, from the face of the earth – men and animals and creatures that move along the ground, and birds of the air – for I am grieved that I have made them. But Noah found favor in the eyes of the

Lord." In Genesis 6:9 (NIV), it goes on to say, "Noah was a righteous man, blameless among the people of his time, and he walked with God."

Because of the wickedness of Noah's generation, God destroyed all of mankind except for Noah, his wife, his three sons, and their wives. The God of the Bible, who does not change, will once again destroy the world because of man's wickedness, depravity, and sinfulness. A prophetic forecast of this event is found in Zephaniah 3:7b-8 (NIV) where we read, "...But they were still eager to act corruptly in all they did. Therefore, wait for me, declares the Lord, for the day I will stand up to testify. I have decided to assemble the nations, to gather the kingdoms and to pour out my wrath on them – all my fierce anger. The whole world will be consumed by the fire of my jealous anger." Additional prophecies concerning the world's destruction can be found in the Book of Revelation, chapters 6-19.

The lesson here is that God is a righteous judge, and because He is righteous, sinning has consequences, even when he has forgiven the sinner. In II Samuel chapter 12, we read the account of Nathan, the prophet, coming to King David and confronting him with the sin of taking Uriah's wife and killing her husband. In II Samuel 12:9b-14 (NIV) Nathan delivers God's condemnation, "You struck down Uriah the Hittite with the sword and took his wife to be your own. Now, therefore, the sword will never depart from your house because you despised me and took the wife of Uriah the Hittite to be your own. This is what the Lord says: "Out of your own household I am going to bring calamity upon you. Before your very eyes, I will take your wives and give them to one who is close to you, and he will lie with your wives in broad daylight. You did it in secret, but I will do this thing in broad daylight before all of Israel. Then David said to Nathan, I have sinned against the Lord. Nathan replied 'The Lord has taken away your sin. You are not going to die. But because by doing this you have made the enemies of the Lord show utter contempt, the son born to you will die.'" Note that the Lord took away David's sin but not the consequences for his sin. All that Nathan prophesied regarding the consequences for his sin came to pass.

The wisdom that comes from the fear of the Lord is intended to keep us from sinning and thereby avoid the resulting consequences. The Bible warns us in Galatians 6:7-8 (NIV), "Do not be deceived: God cannot be mocked. A man reaps what he sows. The one who sows to please the sinful nature, from that nature will reap destruction; the one who sows to please the Spirit, from the Spirit will reap eternal life."

It is God's will, as it was from the beginning, that man not sin. Paul the apostle admonishes the church in Romans 6:1-2 (NIV), "What shall we say, then? Shall we go on sinning so that grace may increase? By no means! We died to sin, how can we live in it any longer?" Finally, one of the most challenging Scriptures in the New Testament is found in I John 3:6 (NIV), which reads, "No one who lives in him keeps on sinning. No one who continues to sin has either seen him or known him." The man who continues to sin, or the country that embraces sinful lifestyles or passes laws contrary to God's commands, will pay a steep price for disobedience. That has been the case throughout history and will continue to be the case until Christ returns.

– 38 –

Two Worldviews and Their Implications

We are presented with two fundamental worldviews: either everything we see around us, from the oceans to the mountains, from the plants to the animals, from the quantum realm to the outer limits of space, is here by random chance and is the product of material and energy, or it is the product of a cosmic designer, God. Your belief regarding these two worldviews will influence your outlook on life and your understanding of your purpose for existence. Either the Big Bang created the universe out of nothing, and there is no divine purpose for being here, or God created the universe and created everyone for a specific purpose. If it is the latter, and the evidence overwhelmingly points in that direction, then there must be a reason for why God did it.

The Bible says that He created man in His image to rule over the earth, and He has a specific purpose for each of us. His desire is to communicate His purpose to us through His Word. He also wants to have a personal relationship with each of us.

The evidence suggests that a purposeful Creator, not an undirected event like the Big Bang, was the true first cause of the universe. If God is the explanation for why the universe appears so precisely fine-tuned to support life, then it follows that He established the laws governing the cosmos, set the exact values for the cosmological constants, and determined the intricate parameters of the anthropic principles. He also designed and encoded the DNA—the intricate instructions for building life itself. Such precision and order point to intentionality, not chance. If this Creator envisioned a universe capable of sustaining human life, He must have done so with purpose, crafting a world that meets all our needs and inspires awe by its incredible beauty.

Beyond providing for physical life, our Creator made us in His image and capable of understanding the complexity of the universe. This suggests He desires more than just our existence—He seeks a relationship with us. However, humanity's introduction of sin disrupted this original harmony, setting in motion the deterioration we see in the world today. Yet, even in a fallen world, the remnants of this divine design remain evident, calling us to reflect on the One who made it all and to repent of our evil ways.

We have disobeyed His instructions and sinned against the God who created us. We were banished from His presence because of disobedience. But in His mercy and unfailing love, He reached out to us while we were yet sinners. He sent prophets to inform us of our errors and to warn us of the consequences of our actions. He sent prophets inspired by the Holy Spirit to write the Bible and to provide us with instructions for living. He has shown us that His Word is unlike any other. It is encrypted with a divine seal and contains prophecies of past and future events. All these things He did for our benefit.

With the information He made available to us, how is it that so many are questioning the authority of His Word? Why are so many believing in theories like abiogenesis (the spontaneous creation of life from non-organic material)? It is because they have been deceived. If so, it is exactly what the Bible predicted. Those who refuse to believe the truth of the Bible are embracing falsehoods, misinformation, and lies as truth.

Regarding the end of the age and God's judgment on the world, we read in Matthew 24:3 (NIV): "As Jesus was sitting on the Mount of Olives, the disciples came to him privately. Tell us, they said, when will this happen and what will be the sign of your coming and of the end of the age? Jesus answered: 'Watch out that no one deceives you.'" One of the greatest signs of the end of the age will be deception! In fact, Jesus said that if it were possible, even the elect would be deceived. However, it won't be possible to deceive the elect since the Holy Spirit and the Word of God guide them. But the masses will be led astray, with propaganda serving as one of Satan's primary tools for widespread deception.

- 39 -

Your Worldview Determines Your Destiny

A wise man once said that your beliefs will determine your actions, and your actions will determine your destiny. The world today is divided between two completely different belief systems or worldviews. One worldview is based on Scripture, which the Holy Spirit gave to the prophets, teachers, and holy men of God. The other is the secular worldview. Those who hold this second view see the world primarily through man's understanding of it, apart from God's revelation. Your choice in how you view the world, i.e., what you choose to believe, will affect every aspect of your life, including politics, religion, science, law, finances, marriage, relationships, morality, ethics, and your understanding of your purpose for being here.

As the laws of logic suggest, both worldviews cannot be true. Either the biblical worldview is true, or the secular worldview is true. Each person must decide for themselves which view they believe. But know for certain that your decision regarding your worldview will affect your approach to every other decision in your life, either consciously or unconsciously. And ultimately, those decisions will determine your future destiny.

Understanding the difference between these two worldviews can sometimes be confusing, especially when a large percentage of the population is attempting to straddle the fence, with one foot in the world and one foot in the Word. For those who find themselves in this camp, consider the words of the apostle John from I John 2:18-19 (NIV), "Dear children, this is the last hour; and as you have heard that the antichrist is coming, even now many antichrists have come. This is how we know it is the last hour. They went out from us, but they did not belong to us. For if they had belonged to us, they would have remained with us; but their going showed that none of them belonged to us." Either Jesus is our Lord, or He isn't. We can't live in both worlds.

The secular world is continually pushing its ideas through television, the internet, social media, institutions of higher learning, politicians, the movie industry, celebrities, and a host of other influencers. As a result, most of the population is being funneled down a pathway that leads to secular humanism, a philosophy based on human reasoning and understanding. Because these influencers are powerful and ubiquitous, a majority of the world's population has moved away from the biblical worldview and embraced secular humanism. Consequently, behaviors and acts clearly defined by the Bible as sinful are being embraced by a large portion of the population. As seen throughout the Scriptures, whenever we rebel against God's commandments, judgment follows. This judgment may manifest as plagues, famine, war, unrighteous rulers or leaders, corrupt governments, and in various other forms.

One of the primary differences that distinguish a secular worldview from a biblical or scriptural worldview is whether one's actions are driven by a desire for personal power, physical pleasure, monetary benefit, self-adulation, or a desire to please God. Bruce Porter, in his YouTube video titled "The Two World Views," discusses freedom of thought and the importance of discrimination and distinction in recognizing eternal truths or differentiating truth from fiction. [196] One must be able to discriminate between values, such as good, better, and best, and distinguish spiritual or biblical morals, such as good or bad, right or wrong, true or false. As stated in the video, "The secular worldview embraces relativism and humanistic teachings which eliminate these attributes through intellectual manipulation. This satanic worldview seeks to blur or even blind our minds to essential and eternal truths until there is no distinction between right and wrong, or between commandments of God and desires of man." [197]

One of the primary goals of secular humanism is to challenge the existence of God by discrediting the Bible and its authority and by turning Scriptural sins into civil rights. When biblical sins are embraced by governments and recharacterized as civil rights, they become law and are accepted by the next generation as morally acceptable. The secular world attempts to replace God's eternal truths and commandments with man's science, morals, opinions, and traditions, using fear and emotion as primary influencers.

Below is a list of questions and answers to help you understand which worldview you embrace and the potential consequences of your choice:

Who am I?

According to Scripture, you are a child of God, made in His image.

According to secular humanism, you are the product of scientific materialism, no different from any other living organism. You just happened to evolve to a higher intelligence because of mutations. These mutations have given you an advantage over other living organisms, but it was just a coincidence of material processes.

Where Did I Come From?

According to the Bible, you are a creation of God. Genesis 1:26 (NIV) states, "And God said, let us make man in our image, after our likeness..." You have been given dominion over the earth and free will to make decisions. The decisions you make will determine your destiny.

The secular worldview says that you came out of a primordial soup and evolved from a single cell, maybe a bacterium, over millions of years. Because you are the product of scientific materialism, all your thoughts and actions are determined by your material composition and life experiences. You really don't have a free will.

Why Am I Here and What is My Purpose?

According to Scripture, you were put here by God to fulfill His purpose. One of your primary life goals should be to seek to understand His purpose for your life.

According to secular humanism, you are simply the product of materialism. You have no divine purpose for being here, and no future destiny other than death and decay.

Where am I going when I die?

According to the Bible, you are a spirit inhabiting a body. Upon the death of your body, your spirit is headed for eternity, either with God or apart from Him. You will either reside in Hell with the devil and the fallen angels or in Heaven with God. Because of His love for you, He has given you the choice of where you will reside.

According to secularism, you are going to the grave. You have no soul and no spirit. There is no future for you other than death, which is the end of your existence. And if this is true, there is no final justice for unpunished crimes, sins, or the lawless acts of humankind toward one another.

Your beliefs regarding the above questions will shape your worldview. They will also motivate your actions and, more importantly, will determine your destiny. The purpose of this book is to provide compelling and convincing evidence of the existence of God and the absolute reliability of the Holy Scriptures.

– 40 –

Rebellion
The World Accepts the Antichrist

The Bible says in 2 Corinthians 4:4 (NIV), "The god of this age has blinded the minds of unbelievers so that they cannot see the light of the gospel of the glory of Christ, who is the image of God." Some may be asking, Who is the god of this age? He is Satan. The Bible describes him in Isaiah 14:13-14 (NIV), "How you have fallen from heaven, O morning star, son of the dawn! You have been cast down to the earth, you who once laid low the nations! You said in your heart, I will ascend to heaven; I will raise my throne above the stars of God; I will sit enthroned on the mount of assembly, on the utmost heights of the sacred mountain. I will ascend above the tops of the clouds; I will make myself like the Most High."

The prophet Ezekiel described Satan in this way in Ezekiel 28:12-14 (NIV), "You were the model of perfection, full of wisdom and perfect in beauty. You were in Eden, the garden of God: every precious stone adorned you… you were anointed as a guardian cherub, for so I ordained you. You were on the holy mount of God; you walked among the fiery stones. You were blameless in your ways from the day you were created till wickedness was found in you."

The Bible says that Satan led a rebellion in heaven, and he and a third of the angels were cast down to the earth. The angelic beings aligned with Satan have one purpose: to lead or influence humans to rebel against God. The apostle Peter provides this warning in I Peter 5:8 (NIV), "Be self-controlled and alert. Your enemy the devil prowls around like a roaring lion looking for someone to devour."

The kingdom of God that Jesus came to establish is being resisted by the devil and his allies. The Bible says that when Jesus returns to the earth, He will defeat the devil and establish his reign for a thousand years. [198] Before that happens, however, a rebellion on Earth will occur. Revelation 13:3b (NIV) says, "The whole world was astonished and followed the beast. Men worshiped the dragon because he had given authority to the beast, and they also worshiped the beast..." Revelation 13:7 (NIV) says, "He was given power to make war against the saints and to conquer them. And he was given authority over every tribe, people, language, and nation."

Finally, in Revelation 20:1-5 (NIV), we read, "...I saw an angel coming down out of heaven, having the key to the Abyss and holding in his hand a great chain. He seized the dragon, that ancient serpent, who is the devil, or Satan, and bound him for a thousand years. He threw him into the Abyss, and locked and sealed it over him, to keep him from deceiving the nations any more until the thousand years were ended. After that, he must be set free for a short time. I saw thrones on which were seated those who had been given authority to judge. And I saw the souls of those who had been beheaded because of their testimony for Jesus and because of the Word of God. They had not worshiped the beast or his image and had not received his mark on their foreheads or their hands. They came to life and reigned with Christ a thousand years. (The rest of the dead did not come to life until the thousand years were ended.) This is the first resurrection."

According to the Bible, the reign of the antichrist will last seven years. It will be a time of great tribulation, the likes of which the world has never seen. It will also be a time of great deception. For the father of lies will deceive the nations and cause them to worship him and give him their allegiance. It says in Matthew 24:24 (NIV), "For false messiahs and false prophets will appear and perform great signs and wonders to deceive, if possible, even the elect."

You may be asking, why will the nations accept the antichrist as their leader? The Bible says in Romans 1:28-32 (NIV), "…since they did not think it worthwhile to retain the knowledge of God, he gave them over to a depraved mind, to do what ought not to be done. They have become filled with every kind of wickedness, evil, greed, and depravity. They are full of envy, murder, strife, deceit, and malice. They are gossips, slanderers, God-haters, insolent, arrogant, and boastful; they invent ways of doing evil; they disobey their parents; they are senseless, faithless, heartless, ruthless. Although they know God's righteous decree that those who do such things deserve death, they not only continue to do these very things but also approve of those who practice them."

- 41 -

End of the Age
Judgment and Tribulation

It appears we have come to this point in history as a direct result of rejecting God and His divine plan for mankind. Because of our refusal to keep His commands, we are fast approaching what the Bible calls the end of the age, a time of judgment and tribulation. This will be followed by a thousand-year period in which Jesus Christ establishes His kingdom on earth. But what proof is there that we're entering the last days, and what does the Bible say about them?

The Bible has much to say about the last days and the end of the present age. The key prophecies concerning this period, the events leading up to it, and Christ's return to the earth are listed below.

In the Book of Matthew 24:3 (NIV), we read, "As Jesus was sitting on the Mount of Olives, the disciples came to him privately. 'Tell us,' they said, 'when will this happen, and what will be the sign of your coming and of the end of the age?'"

In Matthew 24:4 (NIV), Jesus replied, "Watch out that no one deceives you." He knew this period would be a time of great deception, propaganda, and disinformation. Jesus said in Matthew 24:7-8 (NIV), "Nation will rise against nation, and kingdom against kingdom. There will be famines and earthquakes in various places. All these are the beginning of birth pains." At the time of this writing, 27 global conflicts are currently being tracked by the website cfr.org. [199] Reports have suggested that the war in Ukraine is impacting fertilizer supplies and has greatly increased their price, which could potentially lead to global famines. An ISRAEL365NEWS article dated March 7, 2021, suggested that the number and magnitude of earthquakes and volcanoes have been increasing over the past 50 years. [200]

In Matthew 24:9 (NIV), Jesus made the following statement: "... you will be handed over to be persecuted and put to death, and you will be hated by all nations because of me." A recent report by the "Observatory on Intolerance and Discrimination Against Christians" states that Christians in secular environments face more intolerance and discrimination than any other religion worldwide. [201] Because the frequency of persecution against Christians is increasing, perseverance will be necessary.

Regarding the end times, Jesus also said in Matthew 24:10 (NIV), "At that time many will turn away from the faith and will betray and hate each other." A September 2022 article in Christianity Today says the "Decline of Christianity Shows No Signs of Stopping." [202] Jesus went on to say in Matthew 24:12 (NIV), "Because of the increase of wickedness, the love of most will grow cold."

Jesus also told His disciples in Matthew 24:14 (NIV), "And this gospel of the kingdom will be preached in the whole world as a testimony to all nations: and then the end will come." According to Faith in the News, the gospel has been preached to all nations but not to all individuals. [203]

Jesus described signs in the heavens and conditions on Earth during the period of tribulation. In Luke 21:25-26 (NIV) He said, "There will be signs in the sun, moon and stars. On the earth, nations will be in anguish and perplexity at the roaring and tossing of the sea. Men will faint from terror, apprehensive of what is coming on the world, for the heavenly bodies will be shaken." Most of these signs are yet to appear but are prophesied to increase in frequency and intensity once the seven-year tribulation begins.

Jesus concludes his description of the end of the age by telling His disciples how to recognize the period preceding His second coming. In Matthew 24:32-34 (NIV), he said, "Now learn this lesson from the fig tree: As soon as its twigs get tender and its leaves come out, you know that summer is near. Even so, when you see all these things, you know that it is near, right at the door. I tell you the truth, this generation will certainly not pass away until all these things have happened." Jesus is telling us that the generation that sees "all these things" will be the last generation, the one that precedes His second coming and the establishment of His kingdom on earth.

Paul the apostle describes the general frame of mind of people who will be living in the last days. In 2 Timothy 3:1-5 (NIV), he says, "But mark this: There will be terrible times in the last days. People will be lovers of themselves, lovers of money, boastful, proud, abusive, disobedient to their parents, ungrateful, unholy, without love, unforgiving, slanderous, without self-control, brutal, not lovers of the good, treacherous, rash, conceited, lovers of pleasure rather than lovers of God – having a form of godliness but denying its power. Have nothing to do with them."

The above testimonies of Jesus and the apostle Paul provide a detailed description of the specific signs that will announce the coming of the end of the age and the return of Christ. But the keystone, among all the signs, is the rebirth of Israel as a nation. Never in the history of the world has there been a nation that was utterly destroyed and scattered to the ends of the earth that reemerged as a reborn nation. It would have been miraculous indeed for this to have happened by chance, but it was not by chance. The circumstances concerning this nation were all prophesied in God's Word more than two thousand years ago.

God said through Ezekiel the prophet, "...I poured out my wrath on them [Israel] because they had shed blood in the land and because they had defiled it with their idols. I dispersed them among the nations, and they were scattered through the countries; I judged them according to their conduct and their actions." (Ezekiel 36:18-19 NIV). From the time of the Persian Empire's conquest of Israel in 539 BC until the rebirth of the nation on 14 May 1948, the Jewish people were scattered throughout the world. [204]

Ezekiel 36:24 (NIV) says, "For I will take you [Israel] out of the nations; I will gather you from all the countries and bring you back into your own lands." After more than 2,500 years, the Jewish people have returned to their native land. The nation that was previously desolate is now thriving, which was also prophesied by Ezekiel in 36:34-36, (NIV), "The desolate land will be cultivated instead of lying desolate in the sight of all who pass through it. They will say, 'This land that was laid waste has become like the garden of Eden; the cities that were lying in ruins, desolate and destroyed, are now fortified and inhabited.' Then the nations around you that remain will know that I the Lord have rebuilt what was destroyed and have replanted what was desolate. I the Lord have spoken, and I will do it."

The Book of Revelation is the revelation of Jesus to the Apostle John on the Isle of Patmos. It is a prophecy to the churches and the followers of Jesus Christ regarding what is coming at the end of the age. It foretells judgments that are coming upon the earth because of the disobedience of nations and their rejection of God. It tells us of a coming world leader who will install a one-world government and demand allegiance to his rule. Those who do not bow down to his authority and worship him will be executed. He will also establish a new economic system requiring everyone to receive a mark on their right hand or forehead to buy or sell. He will rule for a period of seven years, at the end of which he will be destroyed by the return of the King of kings and the Lord of lords. What follows are a few of the Scriptures that highlight some of these events.

Revelation 13:1 (NIV) says, "And I saw a beast coming out of the sea. He had ten horns and seven heads, with ten crowns on his horns, and on each head a blasphemous name." Revelation 13:3-4 (NIV) says, "One of the heads of the beast seemed to have had a fatal wound, but the fatal wound had been healed. The whole world was astonished and followed the beast. Men worshiped the dragon because he had given authority to the beast, and they also worshiped the beast..." A common belief held by many Christians is that the antichrist will recover from a fatal wound to the head, after which the world will give him their allegiance and ultimately worship him.

Revelation 13:7-8 (NIV) says, "He was given power to make war against the saints and to conquer them. And he was given authority over every tribe, people, language and nation. All inhabitants of the earth will worship the beast – all whose names have not been written in the book of life belonging to the Lamb that was slain from the creation of the world."

Revelation 13:11 (NIV) says, "Then I saw another beast, coming out of the earth. He had two horns like a lamb, but he spoke like a dragon." It goes on to say in verse 13:13,

"And he performed great and miraculous signs, even causing fire to come down from heaven to the earth in full view of men." These signs and miracles will deceive many, causing them to follow the antichrist and the second beast, who is the false prophet.

Revelation 13:15 (NIV) says, "He [the false prophet] was given power to give breath to the image of the first beast, so that it could speak and cause all who refused to worship the image to be killed." Until this generation, it would have been hard to imagine how the world's population could be forced to worship the image of the first beast. But with today's technology, it is now possible to monitor anyone who has a cell phone or TV.

Revelation 13:16 (NIV) goes on to say, "He also forced everyone, small and great, rich and poor, free and slave, to receive a mark on his right hand or on his forehead, so that no one could buy or sell unless he had the mark, which is the name of the beast or the number of his name." Not until this generation did the technology exist that would allow for the creation of an economic system where individuals could buy or sell with a mark on their hand or forehead. His mark might also be a means for monitoring everyone in every nation.

Daniel 9:27 (NIV) tells us, "He [the antichrist] will confirm a covenant with many for one seven, but in the middle of that 'seven,' he will put an end to sacrifice and offering." In other words, he will break the seven-year treaty that he makes with Israel.

What signs should we look for regarding when these things are about to occur? We read in 2 Thessalonians 2:1-4 (NIV), "Concerning the coming of our Lord Jesus Christ and our being gathered to him (the rapture of the church), we ask you, brothers, not to become easily unsettled or alarmed by some prophecy, report or letter supposed to have come from us, saying that the day of the Lord has already come. Don't let anyone deceive you in any way, for that day will not come until the rebellion occurs and the man of lawlessness is revealed, the man doomed to destruction. He opposes and exalts himself over everything that is called God or is worshiped, and even sets himself up in God's temple, proclaiming himself to be God."

Another sign to look for is the rebuilding of the Jewish temple. Most Christians believe Israel will rebuild its temple on the existing Temple Mount in Jerusalem. The construction of the third and final temple will be another sign that we are in the last days.

Chapters six through nineteen of the Book of Revelation describe a time on earth that will be more terrible than any time since the beginning of creation. God's wrath will be poured out on a rebellious and unrepentant population that refuses to acknowledge His lordship or obey His commandments.

The good news for believers is that Christians are not destined for eternal punishment, like the devil and his angels and those who have refused to believe the truth. Hopefully, this book has given you ample reason to believe in the absolute truth of His Word and His plan of salvation. If you have not accepted Jesus Christ as your Lord and Savior, today could be the day of your salvation. Please don't delay your decision; for none of us knows the day or the hour of our departure from this world. If you are ready to receive Him, pray the following prayer in sincerity and I believe He will come into your life, and you will receive His gift of salvation:

Heavenly Father, I come before You today, acknowledging that I am a sinner in need of Your grace. I believe that Jesus Christ is Your Son, that He died on the cross for my sins, and that He rose again to give me eternal life. I repent of my sins and turn away from my old ways, asking You to forgive me and cleanse me. Lord Jesus, I invite You into my heart as my Savior and my Lord. Take control of my life and guide me to live according to Your will. Thank You for loving me, for saving me, and for giving me the gift of eternal life. I trust in You alone for my salvation. In Jesus' name, I pray. Amen.

I leave you with the final words that Jesus spoke to His disciples just before they watched Him ascend into heaven. They are recorded in Mark 16:15-18 (NIV). "Go into all the world and preach the good news to all creation. Whoever believes and is baptized will be saved; but whoever does not believe will be condemned. And these signs will accompany those that believe: In my name they will drive out demons; they will speak in new tongues; they will pick up snakes with their hands; and when they drink deadly poison, it will not hurt them at all; they will place their hands on sick people, and they will get well."

My prayer for you is that His Word will be an everlasting source of good news and that, from this day forward, you will be counted among the believers whose names are written in the Lamb's Book of Life. May God bless and keep you forevermore!

– 42 –

Quotes from the World's Brightest Minds

Sir Isaac Newton – Considered by many to be the greatest scientist who ever lived. He used his binomial theorem to develop calculus and the laws of motion and gravity. Regarding gravity, Newton said, "This most beautiful system of the sun, planets, and comets could only proceed from the counsel and dominion of an intelligent Being. ... This Being governs all things, not as the soul of the world, but as Lord over all; and on account of his dominion, he is wont to be called 'Lord God' παντοκρατωρ [pantokratòr], or 'Universal Ruler'. ... The Supreme God is a Being eternal, infinite, absolutely perfect." [205] Sir Isaac Newton also said, "Atheism is so senseless. When I look at the solar system, I see the earth at the right distance from the sun to receive the proper amounts of heat and light. This did not happen by chance." [206]

Sir Francis Bacon – Lord Chancellor of England and Father of the "Scientific Method." He said, "There are two books laid before us to study, to prevent our falling into error; first, the volume of the Scriptures, which reveal the will of God; then the volume of the Creatures, which express His power." [207] He's also quoted as saying, "A little science estranges a man from God; a lot of science brings him back." [208]

Robert Boyle – Father of modern chemistry. He is best known for describing the inverse relationship between a gas's absolute pressure and volume in a closed system when temperature is kept constant—known as Boyle's Law. [209] He was a devout believer in Jesus Christ and said, "God would not have made the universe as it is unless He intended us to understand it." [210]

John Dalton – Father of Chemistry. He is best known for his atomic theory and color blindness research. He wrote several papers describing gas laws. His law on partial pressure became known as Dalton's Law. [211] Dalton was quoted as saying, "We should scarcely be excused in concluding this essay without calling the reader's attention to the beneficent and wise laws established by the author of nature to provide for the various exigencies of the sublunary creation and to make the several parts dependent upon each other, so as to form one well-regulated system or whole." [212]

Johannes Kepler – Discovered the three mathematical laws of planetary motion, known as Kepler's Laws. These laws explain the discipline of celestial mechanics. He also discovered the elliptical patterns of the planets as they orbit the sun. Kepler made the following profound statements. "Science is the process of thinking God's thoughts after Him." "The chief aim of all investigations of the external world should be to discover the rational order and harmony which has been imposed on it by God and which He revealed to us in the language of mathematics." "[God] is the kind Creator who brought forth nature out of nothing." [213]

Galileo Galilei – Father of modern science. He made major scientific contributions to the fields of physics, astronomy, cosmology, mathematics, and philosophy. [214] Quotes by Galileo include: "To understand the universe, you must understand the language in which it's written, the language of mathematics." "The laws of nature are written by the hand of God in the language of mathematics." [215]

Blaise Pascal – French mathematician, physicist, and philosopher. He is best known for his work in developing probability theory. He is also known for Pascal's Law, which deals with fluid mechanics. [216] He is quoted as saying, "The Christian religion… teaches two truths: that there is a God who men are capable of knowing, and that there is an element of corruption in men that renders them unworthy of God. Knowledge of God without knowledge of man's wretchedness begets pride, and knowledge of man's wretchedness without knowledge of God begets despair, but knowledge of Jesus Christ furnishes man knowledge of both simultaneously." [217]

Gregor Mendel – Father of modern genetics. He discovered the basic principles of heredity through his experiments with pea plants. His observations became the foundation for modern genetics and the laws of heredity. [218] If Mendel's principles of genetics had been widely understood at the time of their discovery, some believe that Darwin's theory of evolution would never have gained widespread acceptance. Since gene pools are finite, they restrict heredity characteristics to those of the progenitors. Therefore, dog genes could not produce cats. It would violate Mendel's laws of heredity. Mendel said, "The victory of Christ gained us the kingdom of grace, the kingdom of heaven." [219]

William Thomson (Lord Kelvin) – Codified the first two laws of thermodynamics and deduced that temperature at absolute zero is -273.15 degrees Celsius. He also invented signaling equipment used in the first transatlantic telegraph. [220] Lord Kelvin is famous for saying, "If you study science deep enough and long enough, it will force you to believe in God." He also said, "Christianity without the cross is nothing. The cross was the fitting close of a life of rejection, scorn and defeat. But in no true sense have these things ceased or changed. Jesus is still He whom man despiseth, and the rejected of men. The world has never admired Jesus, for moral courage is yet needed in every one of its high places by him who would 'confess' Christ. The 'offense' of the cross, therefore, has led men in all ages to endeavor to be rid of it, and to deny that it is the power of God in the world." [221]

Werner Heisenberg – One of the creators of quantum mechanics. He formulated the Heisenberg Uncertainty Principle. [222] Heisenberg made the following declaration: "Where no guiding ideals are left to point the way, the scale of values disappears and with it the meaning of our deeds and sufferings, and at the end can lie only negation and despair. Religion is therefore the foundation of ethics, and ethics the presupposition of life." [223]

Humphry Davy – Discovered the electrical nature of chemical bonding and was the first to split several substances into their basic building blocks using electricity. He also discovered chlorine and iodine and produced the first samples of the following elements: barium, boron, calcium, magnesium, potassium, sodium, and strontium. [224] Quotes by Humphry Davy include the following: "God's design was revealed by chemical investigations." "The three states of the caterpillar, larva, and butterfly have, since the time of the Greek poets, been applied to typify the human being, -- its terrestrial form, apparent death, and ultimate celestial destination." "I envy no quality of the mind or intellect in others; not genius, power, wit, nor fancy; but, if I could choose what would be most delightful, and, I believe, most useful to me, I should prefer a firm religious belief to every other blessing." [225]

Arthur Eddington – First to propose that stars get their energy from nuclear fusion. He also verified Einstein's General Theory of Relativity experimentally. [226] He said, "The idea of a universal mind or Logos would be, I think, a fairly plausible inference from the present state of scientific theory." He also said, "If someone points out to you that your pet theory of the universe is in disagreement with Maxwell's equations – then so much the worse for Maxwell's equations. If it is found to be contradicted by observation – well, these experimentalists do bungle things sometimes. But if your theory is found to be against the second law of thermodynamics, I can give you no hope; there is nothing for it but to collapse in deepest humiliation." [227]

John Ambrose Fleming – Inventor of the vacuum tube, ushering in the electronic age. He developed the hand rules for electric motors and founded the creationist Evolution Protest Movement. [228] His quotes include: "We must not build on the sands of an uncertain and everchanging science...but upon the rock of inspired Scriptures." "The evolution theory is purely the product of the imagination." "Evolution is baseless and quite incredible." "The theory of evolution is totally inadequate to explain the origin and manifestation of the inorganic world." "There is abundant evidence that the Bible, though written by men, is not the product of the human mind. By countless multitudes it has always been revered as a communication to us from the Creator of the Universe." [229]

Samuel Morse – Developed the signaling system known as Morse code and patented the invention of the single-wire telegraph. [230] Morse said, "Education without religion is in danger of substituting wild theories for the simple commonsense rules of Christianity." [231]

John Eccles – Winner of the Nobel Prize in Physiology/Medicine for his work on the physiology of synapses. [232] Eccles said, "I maintain that the human mystery is incredibly demeaned by scientific reductionism, with its claim in promissory materialism to account eventually for all of the spiritual world in terms of patterns of neuronal activity. This belief must be classed as a superstition. ...We have to recognize that we are spiritual beings with souls existing in a spiritual world as well as material beings with bodies and brains existing in a material world." [233]

Gottfried Leibniz – German mathematician, philosopher, physicist, and statesman. He is also known for his independent invention of differential and integral calculus and for his theories about force, energy, and time." [234] Leibniz made the following proclamation: "God, possessing supreme and infinite wisdom, acts in the most perfect manner, not only metaphysically, but also morally speaking, and ... with respect to ourselves, we can say that the more enlightened and informed we are about God's works, the more we will be disposed to find them excellent and in complete conformity with what we might have desired." [235]

James Clerk Maxwell – Best known for demonstrating that light is an electromagnetic wave. He also originated the concept of electromagnetic radiation. His equations, known as Maxwell's equations, helped pave the way for Albert Einstein's special theory of relativity. He is also known for establishing the nature of Saturn's rings and his kinetic theory on gases. "His ideas formed the basis for quantum mechanics and ultimately for the modern theory of the structure of atoms and molecules." [236] Maxwell said, "I have looked into most philosophical systems, and I have seen that none will work without God."

"Science is incompetent to reason upon the creation of matter itself out of nothing. We have reached the utmost limit of our thinking faculties when we have admitted that because matter cannot be eternal and self-existent it must have been created." [237]

James Prescott Joule – "...laid the foundation for the theory of conservation of energy, which later influenced the first law of thermodynamics. He also formulated the Joule's law which deals with the transfer of energy." [238] Quotes by Joule include: "It is evident that an acquaintance with natural laws means no less than an acquaintance with the mind of God therein expressed." "Order is manifestly maintained in the universe...governed by the sovereign will of God." "After the knowledge of, and obedience to, the will of God, the next aim must be to know something of His attributes of wisdom, power and goodness as evidenced by His handiwork." [239]

Joseph Henry – "...was at the forefront of the great electromagnetic advances of the 1830s. He built the world's most powerful electromagnets and made practical breakthroughs, allowing Samuel Morse to invent the telegraph. The unit of electrical inductance is named the *henry* in his honor, with the symbol H. For most of the second half of the 1800s he was America's most renowned scientist." [240] Henry said the following, "The person who thought there could be any real conflict between science and religion must be either very young in science or ignorant of religion." [241]

Leonardo da Vinci – Italian painter, draftsman, sculptor, architect, and engineer. His paintings epitomized the Renaissance era. The Mona Lisa and the Last Supper are two of his most famous paintings. "His notebooks reveal a spirit of scientific inquiry and a mechanical inventiveness that were centuries ahead of their time." [242] Some of his more famous quotes include: "Not to punish evil is equivalent to authorizing it." "It is not enough that you believe what you see. You must also understand what you see." "The greatest deception men suffer is from their own opinions." "There are three classes of people; those who see, those who see when they are shown, those who do not see." [243]

Louis Pasteur – Invented the process of pasteurization, founded microbiology, discovered anaerobic bacteria, revolutionized chemistry and biology, and tied germ theory to the cause of diseases. [244] Louis Pasteur said, "The more I study nature, the more I stand amazed at the work of the Creator. Science brings men nearer to God." "Little science takes you away from God but more of it takes you to Him." "Question your priorities often, then make sure God always comes first." "The greatest derangement of the mind is to believe in something because one wishes it to be so." [245]

Michael Faraday – One of the most influential scientists of the 1800s. "The unit of electrical capacitance is named the *farad* in his honor, with the symbol F." [246] A faraday box, bag, or cage is a device that will protect your cell phone and other electronic equipment from a solar flare or an electromagnetic pulse/radiation (EMP). Notable quotes from Faraday include: "The book of nature which we have to read is written by the finger of God." "Since peace is alone the gift of God, and as it is He who gives it, why should we be afraid? His unspeakable gift in His beloved Son is the ground of no doubtful hope." [247]

Nicolas Copernicus – Polish astronomer and mathematician. He is credited with being the father of modern astronomy. He proposed that the earth and other planets revolve around the sun. Prior to the publication of his famous writings "On the Revolutions of the Heavenly Spheres" in 1543, most astronomers argued that the earth was positioned at the center of the universe. He also correctly postulated the order of planets from the sun and estimated their orbital periods. Copernicus argued that the earth rotated on its axis every 24 hours and gradual shifts accounted for seasonal changes. [248] He stated, "To know the mighty works of God, to comprehend His wisdom and majesty and power; to appreciate, in degree, the wonderful workings of His laws, surely all this must be a pleasing and acceptable mode of worship to the Most High, to whom ignorance cannot be more grateful than knowledge." [249]

Leonhard Euler – Swiss mathematician and physicist. He was responsible for developing many concepts that are integral to modern mathematics and engineering. He also contributed greatly to the fields of geometry, trigonometry, and calculus. Euler was the individual who formulated the equation for the natural logarithm, which is the inverse function of the natural exponential function.

This function is represented by the letter "e" and is called Euler's number. It is a mathematical constant represented by 2.7183..., and like Pi (π), is a nonrepeating number. This brilliant mathematician and physicist said: "Nothing takes place in the world whose meaning is not that of some maximum or minimum." "For since the fabric of the universe is most perfect and the work of a most wise Creator, nothing at all takes place in the universe in which some rule of maximum or minimum does not appear." [250]

Arthur Compton – Received the Nobel Prize in Physics in 1927 for his discovery "that individual scattered X-ray photons and recoil electrons appear at the same instant." [251] "Arthur Compton discovered that light can behave as a particle as well as a wave, and he coined the word *photon* to describe this newly identified particle of light. Compton's discovery was one of the pivotal revelations that led physicists to conclude that objects once thought to be particles can behave like waves and objects once thought to be waves can behave like particles." [252] Compton also played a key role in the development of the first atomic bombs. Notable quotes by Compton include, "For myself, faith begins with a realization that a supreme intelligence brought the universe into being and created man. It is not difficult for me to have this faith, for it is incontrovertible that where there is a plan there is intelligence—an orderly, unfolding universe testifies to the truth of the most majestic statement ever uttered—In the beginning God." "Those whose thinking is disciplined by science, like all others, need a basis for the good life, for aspiration, for courage to do great deeds. They need a faith to live by. The hope of the world lies in those who have such faith and who use the methods of science to make their visions become real. Such visions and hope and faith are not a part of science." [253]

Charles Hard Townes – Co-inventor of the laser and shared the Nobel Prize in Physics in 1964 for his work on the maser-laser principle. He also served as U.S. Presidential Science Advisor and made notable contributions to the field of astrophysics and infrared astronomy. [254] Quotes by Charles Townes follow: "I strongly believe in the existence of God, based on intuition, observations, logic, and also scientific knowledge." "At least this is the way I see it. I am a physicist. I also consider myself a Christian. As I try to understand the nature of our universe in these two modes of thinking, I see many commonalities and crossovers between science and religion. It seems logical that in the long run the two will even converge." "Many have a feeling that somehow intelligence must have been involved in the laws of the universe." [255]

Works Cited

[1] P. Sutter, "Life as we know it would not exist without this highly unusual number," Space.com, 24 March 2022. [Online]. Available: https://www.space.com/fine-structure-constant-universe-mystery. [Accessed 20 November 2024].

[2] "Best-selling book," Guinness World Records, [Online]. Available:https://www.guinnessworldrecords.com/world-records/best-selling-book-of-non-fiction. [Accessed 12 September 2024].

[3] "Holy Bible New International Version," Bible Gateway, [Online]. Available: https://www.biblegateway.com/passage/?search=2+Timothy+3%3A16-17&version=NIV. [Accessed 18 November 2023].

[4] Land Arch Concepts, "The Heptadic Code in Genesis 1:1," [Online]. Available: https://landarchconcepts.wordpress.com/the-heptadic-code-in-genesis-11/. [Accessed 20 August 2023].

[5] Land Arch Concepts, "The Heptadic Code in Genesis 1:1," Land Arch Concepts, [Online]. Available: https://landarchconcepts.wordpress.com/the-heptadic-code-in-genesis-11/. [Accessed 20 August 2023].

[6] Joel, "Genesis 1:1 - Evidence For God," 8 November 2018. [Online]. Available: https://www.christianevidence.net/2018/11/genesis-11-evidence-for-god.html.

[7] C. Hahn, "Why is number 7 considered holy in the Bible? Megachurch pastor explains," Christian Today, 7 April 2016. [Online]. Available: https://www.christiantoday.com/article/why-is-number-7-considered-holy-in-the-bible-megachurch-pastor-explains/83492.htm. [Accessed 22 November 2024].

[8] en.wikipedia.org, "Gematria," Wikipedia The Free Encyclopedia, [Online]. Available: https://en.wikipedia.org/wiki/Gematria. [Accessed 20 August 2023].

[9] I. Panin, "Hepatic structure in the Bible," 8 April 2011. [Online]. Available: https://achristiananswers.blogspot.com/2011/04/hepatic-structure-in-bible.html.

[10] D. C. Missler, "CM," Dr. Chuck Missler's Website - Biography, [Online]. Available: https://chuckmissler.com/biography. [Accessed 25 July 2024].

[11] D. C. Missler, "YouTube: The Last 12 Verses of Mark - Chuck Missler," [Online]. Available: https://www.youtube.com/watch?v=v38aFBB5PlE. [Accessed 20 August 2023].

"Random Fact," factslides.com, 12 January 2025. [Online]. Available: https://www.factslides.com/i-2157#google_vignette. [Accessed 12 January 2025].

[13] D. C. Missler, "YouTube: The Last 12 Verses of Mark Chuck Missler," [Online]. Available: https://www.youtube.com/watch?v=v38aFBB5PlE. [Accessed 20 August 2023].

[14] mymathtables.com, "MYMATHTABLES, "List of First Hundred Hexagonal Numbers," [Online]. Available: https://www.mymathtables.com/numbers/first-hundred-hexagonal-number-table.html.

[15] mymathtables.com, "List of First Hundred Triangular Numbers," MYMATHTABLES, [Online]. Available: https://www.mymathtables.com/numbers/first-hundred-triangular-number-table.html. [Accessed 25 August 2023].

[16] Joel, "Genesis 1:1 - Evidence For God,"

8 November 2018. [Online]. Available: https://www.christianevidence.net/2018/11/genesis-11-evidence-for-god.html.

[17] ChatGPT, "chatgpt.com," ChatGPT Open AI, 10 January 2025. [Online]. Available: https://chatgpt.com/c/67802391-1f18-8011-94f0-ce27b067965d. [Accessed 10 January 2025].

[18] ChatGPT, "chatgpt.com," ChatGPT Open AI, 10 January 2025. [Online]. Available: https://chatgpt.com/c/67802391-1f18-8011-94f0-ce27b067965d. [Accessed 10 January 2025].

[19] Bible Gematria, "The Most Famous Ratio In Mathematics," Bible Gematria, [Online]. Available: https://www.biblegematria.com/pi-and-the-bible.html. [Accessed 29 May 2024].

[20] C. Missler, "The Mysteries of Pi and e - Fundamental Constants?," 1 September 2003. [Online]. Available: https://www.khouse.org/articles/2003/482/.

[21] ChatGPT, "chatgpt.com," ChatGPT Open AI, 13 January 2025. [Online]. Available: https://chatgpt.com/c/67859574-b6d4-8011-b7a9-1b798597554d. [Accessed 13 January 2025].

[22] Joel, "Genesis 1:1 - Evidence For God," Christian Evidence - Evidence That Demands A Verdict, 8 November 2018. [Online]. Available: https://www.christianevidence.net/2018/11/genesis-11-evidence-for-god.html. [Accessed 6 January 2024].

[23] F. C. Payne Revised and Reprinted by E. R. Finck, "The Seal of God, In Creation and the Word, An unanswerable challenge to an unbelieving World", E.R. Finck, 1981.

[24] Joel, "Genesis 1:1 - Evidence For God," Christian Evidence, 8 November 2018. [Online]. Available:

https://www.christianevidence.net/2018/11/genesis-11-evidence-for-god.html. [Accessed 6 January 2024].

[25] "Gematria," Wikipedia, [Online]. Available: https://en.wikipedia.org/wiki/Gematria.
[Accessed 18 February 2024].

[26] I. Gordon, "Bible Study - Resurrection predictions and mysterious numbers!," [Online]. Available: https://jesusplusnothing.com/series/post/bible-study-resurrection-and-numbers-3-8.
[Accessed 12 September 2023].

[27] I. Gordon, "Bible Study - Resurrection predictions and mysterious numbers!," Jesusplusnothing.com, [Online]. Available:https://jesusplusnothing.com/series/post/bible-study-resurrection-and-numbers-3-8.
[Accessed 13 September 2024].

[28] "Genomes. 2nd edition. Chapter 1 The Human Genome," NIH - National Library of Medicine - National Center for Biotechnology Information, [Online]. Available: https://www.ncbi.nlm.nih.gov/books/NBK21134/.
[Accessed 7 January 2024].

[29] S. C. Meyer, Ph.D., Signature In The Cell - DNA and the Evidence for Intelligent Design, HarperCollins, 2009.

[30] J. A. Murugesu, "We now know how many cells there are in the human body," New Scientist (Health), 18 September 2023. [Online]. Available: https://www.newscientist.com/article/2392685-we-now-know-how-many-cells-there-are-in-the-human-body/.
[Accessed 18 February 2024].

[31] S. H. Salomon, "What Is a Protein? A key Building - Block of Life," verywell health, 22 March 2022. [Online]. Available: https://www.verywellhealth.com/what-is-a-protein-5076292.
[Accessed 18 February 2024].

[32] Settlemyer, Latchesar Ionkov and Bradley, "DNA: The Ultimate Data-Storage Solution," 28 May 2021. [Online]. Available: https://www.scientificamerican.com/article/dna-the-ultimate-data-storage-solution/. [Accessed 29 September 2023].

[33] "15.1 The Genetic Code," Openstax, [Online]. Available: https://openstax.org/books/biology-2e/pages/15-1-the-genetic-code. [Accessed 5 March 2024].

[34] S. C. Meyer, Ph.D., Signature in the Cell, DNA and the Evidence for Intelligent Design, New York: HarperCollins, 2009.

[35] S. C. Meyer, Ph.D., Signature in the Cell, DNA and the Evidence for Intelligent Design, New York: HarperCollins, 2009.

[36] Joel, "Genesis 1:1 - Evidence For God," Christian Evidence - Evidence That Demans A Verdict, 8 November 2018. [Online]. Available: https://www.christianevidence.net/2018/11/genesis-11-evidence-for-god.html. [Accessed 13 September 2024].

[37] Joel, "Genesis 1:1 - Evidence For God," Christian Evidence - Evidence That Demands A Verdict, 8 November 2018. [Online]. Available: https://www.christianevidence.net/2018/11/genesis-11-evidence-for-god.html. [Accessed 13 September 2024].

[38] Vladimir I Shcherbak, Ph.D., and Maxim A. Makukov, MS., "The "Wow! signal" of the terrestrial genetic code," Science Direct, 1 May 2013. [Online]. Available: https://www.sciencedirect.com/science/article/abs/pii/S0019103513000791?via%3Dihub. [Accessed 1 August 2024].

[39] Joel, "Genesis 1:1 - Evidence For God," Christian Evidence - Evidence That Demands A Verdict,

8 November 2018 . [Online]. Available: https://www.christianevidence.net/2018/11/genesis-11-evidence-for-god.html. [Accessed 13 September 2024].

[40] Bluer, "Bible Numerics Part 6," [Online]. Available: https://www.youtube.com/watch?v=yT2accYi_ak&t=101s. [Accessed 22 October 2023].

[41] A. i. G. Canada, "We Challenge All Evolutionists to Watch This Video!," [Online]. Available: https://www.youtube.com/watch?v=Dn6i91NRMu8&t=934s. [Accessed 22 October 2023].

[42] News, "Unexplained - Maybe Unexplainable - Numbers Control the Universe," 26 March 2022. [Online]. Available: https://mindmatters.ai/2022/03/unexplained-maybe-unexplainable-numbers-control-the-universe/. [Accessed 23 October 2023].

[43] N. Wolchover, "Physicists Nail Down the 'Magic Number' That Shapes the Universe," 2 December 2020. [Online]. Available: https://www.quantamagazine.org/physicists-measure-the-magic-fine-structure-constant-20201202/. [Accessed 23 October 2023].

[44] N. Wolchover, "Life as We Know It Hinges on One Very Small Decimal," 5 December 2020. [Online]. Available: https://www.theatlantic.com/science/archive/2020/12/the-magic-number-that-shapes-the-universe/617288/. [Accessed 23 October 2023].

[45] P. Sutter, "Life as we know it would not exist without this highly unusual number," Space.com, 24 March 2022. [Online]. Available: https://www.space.com/fine-structure-constant-universe-mystery. [Accessed 7 January 2024].

[46] J. Baez, "How Many Fundamental Constants Are There?," 22 April 2011. [Online]. Available:

https://math.ucr.edu/home/baez/constants.html. [Accessed 23 October 2023].

[47] Elaine Goodfriend, Ph.D., "Seven, the Biblical Number," [Online]. Available: https://www.thetorah.com/article/seven-the-biblical-number. [Accessed 23 October 2023].

[48] "The Fine Structure Constant," Bible Gematria, [Online]. Available: https://www.biblegematria.com/the-mysterious-137.html. [Accessed 25 August 2023].

[49] "The Fine Structure Constant," Bible Gematria, [Online]. Available: https://www.biblegematria.com/the-mysterious-137.html. [Accessed 26 August 2023].

[50] "Fibonacci," Bitannica.com, [Online]. Available: https://www.britannica.com/biography/Fibonacci. [Accessed 7 November 2023].

[51] M. Rini, "How Plants Do Their Math," Physics.aps, 13 June 2013. [Online]. Available: https://physics.aps.org/articles/v6/s85. [Accessed 7 November 2023].

[52] I P Shabalkin, E Yu Grigor'eva, M V Gudkova, P I Shabalkin, "Fibonacci Sequence and Supramolecular Structure of DNA," NIH National Library of Medicine - Pub Med, 6 June 2016. [Online]. Available: https://pubmed.ncbi.nlm.nih.gov/27265133/. [Accessed 7 November 2023].

[53] Michel E Beleza Yamagishi, Alex Itiro Shimabukuro, "Nucleotide frequencies in human genome and fibonacci numbers," NIH National Library of Medicine National Center for Biotechnology Information - Pub Med, 10 November 2007. [Online]. Available: https://pubmed.ncbi.nlm.nih.gov/17994268/. [Accessed 7 November 2023].

[54] G. Meisner, "DNA spiral as a Golden Section," Phi 1.618 The Golden Number, 13 May 2012. [Online].

Available: https://www.goldennumber.net/dna/. [Accessed 7 November 2023].

[55] J. Manangan, "The Fibonacci Sequence in Nature," coe.hawaii.edu, [Online]. Available: https://coe.hawaii.edu/ethnomath/wp-content/uploads/sites/12/2019/10/Fibonacci-Sequence-in-Nature.pdf. [Accessed 7 November 2023].

[56] J. Manangan, "The Fibonacci Sequence in Nature," coe.hawaii.edu, [Online]. Available: https://coe.hawaii.edu/ethnomath/wp-content/uploads/sites/12/2019/10/Fibonacci-Sequence-in-Nature.pdf. [Accessed 7 November 2023].

[57] cosmicarchives, "Gods Fingerprint - The Fibonacci Sequence - Golden Ratio and the Fractal Nature of Reality," Cosmic Archives, [Online]. Available: https://www.youtube.com/watch?v=4VrcO6JaMrM&t=271s. [Accessed 12 December 2023].

[58] R. M. Skobac, "The Name of God and Why Jews Don't Say GOD's Name? - Rabbi Michael Skobac," Jews for Judaism , [Online]. Available: https://www.youtube.com/watch?v=sx_6kOrv8Bg. [Accessed 7 November 2023].

[59] Dr. C. Missler, "Chuck Missler: Adam - Noah; The Genealogy, The Translation and the Prophecy. Part 1," YouTube, [Online]. Available: https://www.youtube.com/watch?v=tSNeGqoYb_8. [Accessed 3 August 2024].

[60] "Welcome to the Shroud of Turin Website," Shroud.com, 6 July 2023. [Online]. Available: https://www.shroud.com/menu.htm. [Accessed 8 November 2023].

[61] "Why Pollen on the Shroud of Turin Proves it is Real," Early Church History.org, 5 September 2018. [Online]. Available: https://earlychurchhistory.org/christian-

symbols/why-pollen-on-the-shroud-of-turin-proves-it-is-real/. [Accessed 8 November 2023].

[62] Daniel Esparza, "Shroud of Turin coins may finally have been identified," Aleteia.com, 26 April 2017. [Online]. Available: https://aleteia.org/2017/04/26/shroud-of-turin-coins-may-finally-have-been-identified/. [Accessed 8 November 2023].

[63] "STURP Investigations 1978 - Shroud of Turin Research Project," Shroud 3D, [Online]. Available: https://shroud3d.com/addendum/sturp-1978/. [Accessed 8 November 2023].

[64] Ray Downing - 3D Illustrator and Animator, "The 3D Information on the Shroud of Turin," Ray Downing, 16 February 2016. [Online]. Available: https://www.raydowning.com/blog/2016/2/15/the-3d-information-on-the-shroud-of-turin. [Accessed 8 November 2023].

[65] B. Yirka, "Study of data from 1988 Shroud of Turin testing suggests mistakes," Phys.org, [Online]. Available: https://phys.org/news/2019-07-shroud-turin.html#:~:text=In%20this%20new%20effort%2C%20the,shroud—just%20some%20edge%20pieces.. [Accessed 8 November 2023].

[66] M. Molac, "Turin Shroud - Cotton adulterated 1988 Carbon 14 Test," marius molac , [Online]. Available: https://www.youtube.com/watch?v=51RrDewZqos. [Accessed 8 November 2023].

[67] "The Shroud of Turin Proof of Authenticity Beyond Reasonable Doubt (1 of 2)," rainbowlightstudio, [Online]. Available: https://www.youtube.com/watch?v=sJymwctqo-A&t=435s. [Accessed 8 November 2023].

[68] T. &. E. Wayland, "Italian Scientist Says New X-Ray Dating Technique Shows Turin Shroud to be 2000 Years

Old," Singular Fortean, 30 April 2022. [Online]. Available: https://www.singularfortean.com/news/2022/4/29/italian-scientist-says-new-x-ray-dating-technique-shows-turin-shroud-to-be-2000-years-old. [Accessed 8 November 2023].

[69] "The Shroud of Turin: Proof of Authenticity Beyond Reasonable Doubt (1 of 2)," rainbowlightstudio, [Online]. Available: https://www.youtube.com/watch?v=sJymwctqo-A. [Accessed 8 November 2023].

[70] chatGPT, "chatgpt.com," ChatGPT by OpenAI, 2 December 2024. [Online]. Available: https://chatgpt.com/c/674ccbf8-d178-8011-bd80-13e263d1517d. [Accessed 2 December 2024].

[71] "How Image was formed on the Shroud of Turin FAST VERSION," Good Sheperd Films, [Online]. Available: https://www.youtube.com/watch?v=QkmmEjfsNG4&t=437s. [Accessed 8 November 2023].

[72] Dr. C. Missler, "Confirming the Prophetic Date of 445 B.C.," Koinonia House - Book "The Creator Beyond Time and Space", 9 January 2025. [Online]. Available: http://www.xwalk.ca/king2.html. [Accessed 9 January 2025].

[73] ChatGPT, "ChatGPT," ChatGPT Open AI, 9 January 2025. [Online]. Available: https://chatgpt.com/c/67802ae9-e25c-8011-9093-52ec70254865. [Accessed 9 January 2025].

[74] BEC CREW, "Scientists Just Captured The Flash of Light That Sparks When a Sperm Meets an Egg," Science alert, 27 April 2016. [Online]. Available: https://www.sciencealert.com/scientists-just-captured-the-actual-flash-of-light-that-sparks-when-sperm-meets-an-egg. [Accessed 18 February 2024].

[75] BEC CREW, "Scientists Just Captured The Flash of Light That Sparks When a Sperm Meets an Egg,"

27 April 2016. [Online]. Available: https://www.sciencealert.com/scientists-just-captured-the-actual-flash-of-light-that-sparks-when-sperm-meets-an-egg. [Accessed 18 February 2024].

[76] "Georges Lemaitre, Father of the Big Bang," American Museum of Natural History, [Online]. Available: https://www.amnh.org/learn-teach/curriculum-collections/cosmic-horizons-book/georges-lemaitre-big-bang. [Accessed 4 November 2023].

[77] "Time - STARS Where Life Begins," Time Magazine, 27 December 1976. [Online]. Available: https://content.time.com/time/subscriber/article/0,33009,947769-3,00.html. [Accessed 4 November 2023].

[78] "Which star / galaxy is moving away from us the fastest?," Astronomy, [Online]. Available: https://astronomy.stackexchange.com/questions/32582/which-star-galaxy-is-moving-away-from-us-the-fastest. [Accessed 4 November 2023].

[79] R. R. Britt, "Huge Hole Found in the Universe," Space.com, 23 August 2007. [Online]. Available: https://www.space.com/4271-huge-hole-universe.html. [Accessed 4 November 2023].

[80] A. N. Meldrum, "The Discovery of the Weight of the Air," Nature, 30 July 1908. [Online]. Available: https://www.nature.com/articles/078294a0. [Accessed 4 November 2023].

[81] B. Jackson, "Astro for kids: How many stars are there in space?," Astronomy, 28 September 2021. [Online]. Available: https://www.astronomy.com/science/astro-for-kids-how-many-stars-are-there-in-space/. [Accessed 4 November 2023].

[82] D. Djurisic, "Nothing Is Solid & Everything Is Energy - Scientists Explain The World of Quantum Physics*," Linked in, 7 January 2016. [Online]. Available: https://www.linkedin.com/pulse/nothing-solid-everything-energy-scientists-explain-world-djurisic. [Accessed 4 November 2023].

[83] "StarChild Question of the Month for February 2000 - Does the Sun move around the Milky Way?," Star Child, February 2000. [Online]. Available: https://starchild.gsfc.nasa.gov/docs/StarChild/questions/question18.html. [Accessed 5 November 2023].

[84] "Antibacterial and antifungal activities of thymol: A brief review of the literature," Science Direct, 1 November 2016. [Online]. Available: https://www.sciencedirect.com/science/article/abs/pii/S0308814616306392. [Accessed 5 November 2023].

[85] "Why is a male child circumcised on the 8th day?," Biblical Hermeneutics, [Online]. Available: https://hermeneutics.stackexchange.com/questions/31778/why-is-a-male-child-circumcised-on-the-8th-day. [Accessed 5 November 2023].

[86] "Do goats chew cud," Answers, 10 August 2023. [Online]. Available: https://www.answers.com/zoology/Do_goats_chew_cud. [Accessed 5 November 2023].

[87] "Heart Intelligence," HeartMath Institute, 7 August 2012. [Online]. Available: https://www.heartmath.org/articles-of-the-heart/the-math-of-heartmath/heart-intelligence/. [Accessed 5 November 2023].

[88] "Heart Intelligence," HeartMath Institute, 7 August 2012. [Online]. Available: https://www.heartmath.org/articles-of-the-heart/the-math-of-heartmath/heart-intelligence/. [Accessed 5 November 2023].

[89] S. Joshi, "Memory transference in organ transplant recipients," Journal of New Approaches to Medicine and Health NAMAH, 24 April 2011. [Online]. Available: http://www.namahjournal.com/doc/Actual/Memory-transference-in-organ-transplant-recipients-vol-19-iss-1.html. [Accessed 5 November 2023].

[90] S. Joshi, "Memory transference in organ transplant recipients," Journal of New Approaches to Medicine and Health, 24 April 2011. [Online]. Available: http://www.namahjournal.com/doc/Actual/Memory-transference-in-organ-transplant-recipients-vol-19-iss-1.html. [Accessed 5 November 2023].

[91] "Hebrew numerals," Wikipedia, [Online]. Available: https://en.wikipedia.org/wiki/Hebrew_numerals. [Accessed 25 November 2023].

[92] K. Sabiers, Mathematics Prove Holy Scriptures, Tell International, 1969.

[93] J. Seegert, "Jay Seegert: Christianity, Logic and Science," [Online]. Available: https://www.youtube.com/watch?v=kzC-wBRMfx4. [Accessed 24 October 2023].

[94] M. A. Francisco, "History of Crucifixion Explained," Gunge, Updated 28 February 2023. [Online]. Available: https://www.grunge.com/587858/history-of-crucifixion-explained/. [Accessed 7 January 2024].

[95] "How many prophecies did Jesus fulfill?," Got Questions, [Online]. Available: https://www.gotquestions.org/prophecies-of-Jesus.html. [Accessed 25 November 2023].

[96] R. N. Frye, "Cyrus the Great," Britannica, 9 October 2023. [Online]. Available: https://www.britannica.com/biography/Cyrus-the-Great. [Accessed 24 October 2023].

[97] J. Smith, "2 Chronicles," Bible Hub, [Online]. Available: https://biblehub.com/summary/2_chronicles/1.htm. [Accessed 3 August 2024].

[98] H. Adams, "Who Was Ezra and Why Is His Book Significant?," Bible Study Tools, 23 October 2023. [Online]. Available: https://www.biblestudytools.com/bible-study/topical-studies/who-was-ezra-and-why-is-his-book-significant.html. [Accessed 3 August 2024].

[99] "logic," Merriam Webster Dictionary, [Online]. Available: https://www.merriam-webster.com/dictionary/logic. [Accessed 25 October 2023].

[100] "reason," Merriam Webster Dictionary, [Online]. Available: https://www.merriam-webster.com/dictionary/reason. [Accessed 25 October 2023].

[101] "Laws of Thought," Britannica, [Online]. Available: https://www.britannica.com/topic/laws-of-thought. [Accessed 25 October 2023].

[102] "Inorganic - Definition," Biology Online, [Online]. Available: https://www.biologyonline.com/dictionary/inorganic. [Accessed 26 October 2023].

[103] "Wordnik," The Atlantic Heritage Dictionary of the English Language 5th Edition, [Online]. Available: https://www.wordnik.com/words/organic. [Accessed 19 January 2025].

[104] T. Hughbanks, "Entropy and 2nd Law of Thermodynamics CHEM 102," Texas A&M University, [Online]. Available: https://www.chem.tamu.edu/rgroup/hughbanks/courses/102/slides/slides3_2.pdf. [Accessed 13 August 2024].

[105] "What the First Two Laws of Thermodynamics Are and Why They Matter," Interesting Engineering, [Online]. Available: https://interestingengineering.com/science/what-the-first-two-laws-of-thermodynamics-are-and-why-they-matter. [Accessed 26 October 2023].

[106] "Entropy," Merriam Webster Dictionary, [Online]. Available: https://www.merriam-webster.com/dictionary/entropy. [Accessed 26 October 2023].

[107] "The Big Bang Conundrum: Unraveling JWST's Mystifying Findings on the Early Universe," Science Time, [Online]. Available: https://www.youtube.com/watch?v=TudUD97g4YY. [Accessed 10 January 2024].

[108] Rhett Herman and Tsunefumi Tanaka, "What causes objects such as stars and black holes to spin?," Scientific America, 1 March 1999. [Online]. Available: https://www.scientificamerican.com/article/ what-causes-objects-such/. [Accessed 29 November 2023].

[109] Frank Turek, PH.D., Norman L. Geisler, PH. D., I Don't Have Enough Faith to Be an Atheist, Wheaton, Illinois: Crossway Books, 2004.

[110] "What is the Anthropic Principle?," Got Questions - Your Questions. Biblical Answers, [Online]. Available: https://www.gotquestions.org/anthropic-principle.html. [Accessed 3 November 2023].

[111] "anthropic principle," Britannica, 24 October 2023. [Online]. Available: https://www.britannica.com/science/anthropic-principle. [Accessed 29 November 2023].

[112] M. McFall-Johnsen, "Earth is screaming through space at 1.3 million mph. A simple animation by a former NASA scientist shows what that looks like.," Business

Insider, [Online]. Available: https://www.businessinsider.com/earth-screaming-through-space-nasa-animated-video-2019-10?op=1. [Accessed 3 November 2023].

[113] S. Youngren, "OK...I want numbers. What is the probability the universe is the result of chance?," God Evidence, 1 December 2010. [Online]. Available: https://godevidence.com/2010/12/ok-i-want-numbers-what-is-the-probability-the-universe-is-the-result-of-chance/. [Accessed 3 November 2023].

[114] S. Youngren, "OK...I want numbers. What is the probability the universe is the result of chance?," God Evidence, 1 December 2010. [Online]. Available: https://godevidence.com/2010/12/ok-i-want-numbers-what-is-the-probability-the-universe-is-the-result-of-chance/. [Accessed 3 November 2023].

[115] J. Toupos, "Our Spooky Universe: Fine-Tuning and God," Amos37, [Online]. Available: https://amos37.com/anthropic-constants/. [Accessed 3 November 2023].

[116] ChatGPT, "ChatGPT.com," ChatGPT OpenAI, 3 December 2024. [Online]. Available: https://chatgpt.com/c/674dc0db-dcf8-8011-ae0b-c2d6c0f0d3dd. [Accessed 3 December 2024].

[117] R. Collins, "The Fine-Tuning Design Argument - A Scientific Argument for the Existence of God," Discovery Institute, 1 September 1998. [Online]. Available: https://www.discovery.org/a/91/. [Accessed 9 December 2024].

[118] J. Cafasso, "How Many Cells Are in the Human Body? Fast Facts," Healthline.com, 18 July Updated 2018. [Online]. Available: https://www.healthline.com/health/number-of-cells-in-body. [Accessed 3 November 2023].

[119] M. J. Lopez and S. S. mohiuddin, "Biochemistry, Essential Amino Acids," National Library of Medicine - National Center for Biotechnology Information, 13 March 2023. [Online]. Available: https://www.ncbi.nlm.nih.gov/books/NBK557845/. [Accessed 30 November 2023].

[120] "3.7: Proteins - Types and Functions of Proteins," LibreTexts Biology, [Online]. Available: https://bio.libretexts.org/Bookshelves/Introductory_and_General_Biology/Book%3A_General_Biology_(Boundless)/03%3A_Biological_Macromolecules/3.07%3A_Proteins_-_Types_and_Functions_of_Proteins. [Accessed 30 November 2023].

[121] "Protein," Wikipedia, [Online]. Available: https://en.wikipedia.org/wiki/Protein. [Accessed 4 November 2023].

[122] D. Goodsell, "Molecule of the Month: Titin," PDB-101, May 2015. [Online]. Available: https://pdb101.rcsb.org/motm/185. [Accessed 22 August 2024].

[123] G. Easterbrook, "Before the Big Bang," *U.S. News & World Report, special edition,* p. 16, 2003.

[124] H. Ross, Ph.D., "Anthropic Principle: A Precise Plan for Humanity," Reasons.org, 1 January 2002. [Online]. Available: https://reasons.org/explore/publications/facts-for-faith/anthropic-principle-a-precise-plan-for-humanity. [Accessed 2023 November 2023].

[125] J. Stockwell, "Borel's Law and the Origin of Many Creationist Probability Assertions," The TalkOrigins Archive - Exploring the Creation/Evolution Controversy, 13 March 2002. [Online]. Available: https://talkorigins.org/faqs/abioprob/borelfaq.html. [Accessed 21 January 2025].

[126] P. b. J. Toupos, "Our Spooky Universe: Fine-Tuning and God," Amos37, [Online]. Available: https://amos37.com/anthropic-constants/. [Accessed 3 November 2023].

[127] Casey Luskin, Ph.D., J.D., M.S. "The Top Six Lines of Evidence for Intelligent Design," Discovery Institute , 25 February 2021. [Online]. Available: https://www.discovery.org/a/sixfold-evidence-for-intelligent-design/. [Accessed 21 August 2024].

[128] "Anthropic principle," New World Encyclopedia, [Online]. Available: https://www.newworldencyclopedia.org/entry/Anthropic_principle#google_vignette. [Accessed 21 August 2024].

[129] D. H. Bailey, Math Scholar, 22 November 2017. [Online]. Available: https://mathscholar.org/2017/11/fine-tuning-and-fermis-paradox/. [Accessed 21 August 2024].

[130] "Cosmic Fine Tuning," All About Science, [Online]. Available: https://www.allaboutscience.org/cosmic-fine-tuning.htm. [Accessed 18 September 2024].

[131] "Cosmic Fine Tuning," All About Science, [Online]. Available: https://www.allaboutscience.org/cosmic-fine-tuning.htm. [Accessed 21 August 2024].

[132] "Cosmic Fine Tuning," All About Science, [Online]. Available: https://www.allaboutscience.org/cosmic-fine-tuning.htm. [Accessed 21 August 2024].

[133] J. W. Richards, "List of Fine-Tuning Parameters," Discovery Institute, 14 January 2015. [Online]. Available: https://www.discovery.org/a/fine-tuning-parameters/. [Accessed 21 January 2025].

[134] Jay Richards, Ph.D., "List of Fine-Tuning Parameters," Intelligent Design, [Online]. Available: https://intelligentdesign.org/articles/list-of-fine-tuning-parameters/. [Accessed 21 August 2024].

[135] K. J. Brooks, "What are the odds of winning the Powerball jackpot?," CBS News, 7 November 2022. [Online]. Available: https://www.cbsnews.com/news/powerball-jackpot-record-amount-19-billion-odds-of-winning/. [Accessed 29 November 2023].

[136] ChatGPT, "chatgpt.com," ChatGPT OpenAI, 7 December 2024. [Online]. Available: https://chatgpt.com/c/6753ad01-8ff8-8011-a248-4dcc7a80bf7e. [Accessed 7 December 2024].

[137] ChatGPT, "chatgpt.com," ChatGPT OpenAI, 7 December 2024. [Online]. Available: https://chatgpt.com/c/6753ad01-8ff8-8011-a248-4dcc7a80bf7e. [Accessed 7 December 2024].

[138] Dr. Gary Parker, "2.5 Mutations, Yes; Evolution, No," Answers in Genesis, [Online]. Available: https://answersingenesis.org/genetics/mutations/mutations-yes-evolution-no/. [Accessed 10 January 2024].

[139] University of Toronto, "A simple cell holds 43 million protein molecules, scientists reveal," PHYS.ORG, 17 January 2018. [Online]. Available: https://phys.org/news/2018-01-simple-cell-million-protein-molecules.html. [Accessed 26 October 2023].

[140] B10NUMB3R5, "Size of average protein," B10NUMB3R5, [Online]. Available: https://bionumbers.hms.harvard.edu/bionumber.aspx?id=105224&ver=12. [Accessed 26 October 2023].

[141] G. P. Collins, "Claude E. Shannon: Founder of Information Theory," Scientific American - Technology, 14 October 2002. [Online]. Available: https://www.scientificamerican.com/article/claude-e-shannon-founder/. [Accessed 27 October 2023].

[142] Rob Goodman and Jimmy Soni, "Claude Shannon: The Juggling Poet Who Gave Us the Information Age," Daily

Beast, 30 July 2017. [Online]. Available: https://www.thedailybeast.com/claude-shannon-the-juggling-poet-who-gave-us-the-information-age. [Accessed 2 March 2024].

[143] P. C, "Origin: Probability of a Single Protein Forming by Chance," Philip C, [Online]. Available: https://www.youtube.com/watch?v=W1_KEVaCyaA. [Accessed 28 November 2023].

[144] "Professor Werner Gitt's Conclusions from the Information found in DNA," Beyond Today, 22 May 2005. [Online]. Available: https://www.ucg.org/the-good-news/professor-werner-gitts-conclusions-from-the-information-found-in-dna. [Accessed 31 October 2023].

[145] "Meaning of Information," Treasure Words, [Online]. Available: https://www.treasurewords.com/meaning-of-information.html. [Accessed 31 October 2023].

[146] Bradley Settlemyer and Latchesar Ionkov, "DNA: The Ultimate Data-Storage Solution," Scientific America, 28 May 2021. [Online]. Available: https://www.scientificamerican.com/article/dna-the-ultimate-data-storage-solution/. [Accessed 31 October 2023].

[147] "Which came first, the chicken or the egg?," New Scientist, [Online]. Available: https://www.newscientist.com/question/came-first-chicken-egg/. [Accessed 1 March 2024].

[148] "Intelligent Design," American Heritage Dictionary, [Online]. Available: https://www.ahdictionary.com/word/search.html?q=intelligent+design&submit.x=51&submit.y=19. [Accessed 3 November 2023].

[149] F. C. Payne Revised and Reprinted by E. R. Finck,

"The Seal of God, In Creation and the Word, An unanswerable challenge to an unbelieving World", E.R. Finck, 1981.

[150] R. Hooper, "Hugh Everett: The man who gave us the multiverse," New Scientist, 24 September 2014. [Online]. Available: https://www.newscientist.com/article/dn26261-hugh-everett-the-man-who-gave-us-the-multiverse/. [Accessed 3 November 2023].

[151] "Multiverse," Britannica, [Online]. Available: https://www.britannica.com/science/multiverse. [Accessed 3 November 2023].

[152] S. C. Jamie Carter, "Is The Multiverse Real? The Science Behind 'Everythng Everywhere All At Once'," Forbes, 12 March 2023. [Online]. Available: https://www.forbes.com/sites/jamiecartereurope/2023/03/12/is-the-multiverse-real-the-science-behind-everything-everywhere-all-at-once/?sh=2cc266534754. [Accessed 3 November 2023].

[153] S. C. Jamie Carter, "Is The Multiverse Real? The Science Behind 'Everything Everywhere All At Once'," Forbes, 12 March 2023. [Online]. Available: https://www.forbes.com/sites/jamiecartereurope/2023/03/12/is-the-multiverse-real-the-science-behind-everything-everywhere-all-at-once/?sh=2cc266534754. [Accessed 3 November 2023].

[154] P. J. Kiger, "How Old Is Earth an How Did Scientists Figure It Out?," howstuffworks, 4 February 2021. [Online]. Available: https://science.howstuffworks.com/how-old-is-earth.htm. [Accessed 30 November 2023].

[155] "How Long Have Humans Been On Earth?," World Atlas, [Online]. Available: https://www.worldatlas.com/articles/how-long-have-humans-been-on-earth.html.

[Accessed 5 November 2023].

[156] "Nebraska Man," Wikipedia, [Online]. Available: https://en.wikipedia.org/wiki/Nebraska_Man. [Accessed 5 November 2023].

[157] "Nebraska Man," Wikipedia, [Online]. Available: https://en.wikipedia.org/wiki/Nebraska_Man. [Accessed 5 November 2023].

[158] "Java Man," Wikipedia, [Online]. Available: https://en.wikipedia.org/wiki/Java_Man. [Accessed 5 November 2023].

[159] F. C. Payne Revised and Reprinted by E. R. Finck, "The Seal of God, In Creation and the Word, An unanswerable challenge to an unbelieving World", E.R. Finck, 1981.

[160] "Java Man," Wikipedia, [Online]. Available: https://en.wikipedia.org/wiki/Java_Man. [Accessed 5 November 2023].

[161] "Homo heidelbergensis," Britannica.com, [Online]. Available: https://www.britannica.com/topic/Homo-heidelbergensis. [Accessed 5 November 2023].

[162] F. C. Payne Revised and Reprinted by E. R. Finck, "The Seal of God, In Creation and the Word, An unanswerable challenge to an unbelieving World", E.R. Finck, 1981.

[163] "Neanderthal," Wikipedia, [Online]. Available: https://en.wikipedia.org/wiki/Neanderthal. [Accessed 5 November 2023].

[164] F. C. Payne Revised and Reprinted by E. R. Finck, "The Seal of God, In Creation and the Word, An unanswerable challenge to an unbelieving World", E.R. Finck, 1981.

[165] "Piltdown Man," Britannica.com, [Online]. Available: https://www.britannica.com/topic/Piltdown-man. [Accessed 5 November 2023].

[166] F. C. Payne Revised and Reprinted by E. R. Finck, "The Seal of God, In Creation and the Word, An unanswerable challenge to an unbelieving World", E.R. Finck, 1981.

[167] "The Age of the Earth with Dr Russell Humphreys," Educate For Life Video, [Online]. Available: https://www.youtube.com/watch?v=2QjlAPNC16M. [Accessed 19 January 2025].

[168] Jeffrey P. Tomkins, Ph.D., "Six Biological Evidences for a Young Earth," Institute for Creation Research - ACTS & FACTS IMPACT, 30 APRIL 2019. [Online]. Available: https://www.icr.org/article/six-biological-evidences-for-a-young-earth/. [Accessed 6 November 2023].

[169] Jeffrey P. Tomkins, Ph.D., "Six Biological Evidences for a Young Earth," Institute for Creation Research - ACTS & FACTS IMPACT, 30 April 2019. [Online]. Available: https://www.icr.org/article/six-biological-evidences-for-a-young-earth/. [Accessed 6 November 2023].

[170] Robert W. Carter, Stephen Lee, and John C. Sanford, "An Overview of the Independent Histories of the Human Y Chromosome and the Human Mitochondrial Chromosome," iCC The Proceedings of the International Conference on Creation, 2018. [Online]. Available: https://digitalcommons.cedarville.edu/cgi/viewcontent.cgi?article=1082&context=icc_proceedings. [Accessed 6 November 2023].

[171] Robert W. Carter, Stephen Lee, and John C. Sanford, "An Overview of the Independent Histories of the Human Y Chromosome and the Human Mitochondrial Chromosome," Proceedings of the Eighth International Conference of Creationism, 2018. [Online]. Available: https://digitalcommons.cedarville.edu/cgi/viewcontent.cgi?article=1082&context=icc_proceedings. [Accessed 8 December 2023].

[172] Brian Thomas, M.S., "Published Reports of Original Soft Tissue Fossils," Institute for Creation Research, 17 September 2018. [Online]. Available: https://www.icr.org/soft-tissue-list/. [Accessed 6 November 2023].

[173] Jeffrey P. Tomkins, Ph.D., "Six Biological Evidences for a Young Earth," Institute for Creation Research, 30 April 2019. [Online]. Available: https://www.icr.org/article/six-biological-evidences-for-a-young-earth/. [Accessed 6 November 2023].

[174] Jeffrey P. Tomkins, Ph.D., "Six Biological Evidences for a Young Earth," Institute for Creation Research, 30 April 2019. [Online]. Available: https://www.icr.org/article/six-biological-evidences-for-a-young-earth/. [Accessed 8 December 2023].

[175] Jeffrey P. Tomkins, Ph.D. and Timothy L. Clarey,Ph.D., "Red Algae Lazarus Effect Can't Resurrect Evolution," Institute for Creation Research, 31 January 2019. [Online]. Available: https://www.icr.org/article/red-algae-lazarus-effect-cant-resurrect-evolution/. [Accessed 6 November 2023].

[176] Robert Carter and Chris Hardy, "Modelling biblical human population growth," Creation.com, April 2015. [Online]. Available: https://creation.com/biblical-human-population-growth-model. [Accessed 6 November 2023].

[177] Brian Thomas, Ph.D., "DNA in Dinosaur Bones?," Institute for Creation Research - ACTS & FACTS IMPACT, 28 December 2012. [Online]. Available: https://www.icr.org/article/7160/. [Accessed 6 November 2023].

[178] Brian Thomas, Ph.D., "DNA in Dinosaur Bones?," Institute for Creation Research ACTS & FACTS IMPACT,

28 December 2012. [Online]. Available: https://www.icr.org/article/7160/. [Accessed 6 November 2023].

[179] "carbon-14," Britannica.com, [Online]. Available: https://www.britannica.com/science/carbon-14. [Accessed 6 November 2023].

[180] Dr. Andrew A. Snelling, "Carbon-14 in Fossils and Diamonds," Answers in Genesis, [Online]. Available: https://answersingenesis.org/geology/carbon-14/carbon-14-in-fossils-and-diamonds/. [Accessed 6 November 2023].

[181] "Space Place Explore Earth and Space," NASA Science, 20 December 2021. [Online]. Available: https://spaceplace.nasa.gov/comets/en/. [Accessed 17 August 2024].

[182] Jason Lisle, Ph.D., "The Solar System: Asteroids and Comets," Institute for Creation Research - ACTS & FACTS IMPACT, 30 April 2014. [Online]. Available: https://www.icr.org/article/solar-system-asteroids-comets. [Accessed 6 November 2023].

[183] Andrew A. Snelling, Ph.D. and David E. Rush, M.S., "Moon Dust and the Age of the Solar System," Institute for Creation Research, 1 April 1992. [Online]. Available: http://static.icr.org/i/pdf/technical/Moon-Dust-and-the-Age-of-the-Solar-System.pdf. [Accessed 6 November 2023].

[184] John D. Morris, Ph.D. "Earth's Magnetic Field," Institute for Creation Research - ACTS & FACTS IMPACT, 1 August 2010. [Online]. Available: https://www.icr.org/article/earths-magnetic-field/. [Accessed 6 November 2023].

[185] Russell Humphreys, Ph.D. "The Creation of Cosmic Magnetic Fields," Institute for Creation Research - ACTS & FACTS IMPACTS, 3 August 2008. [Online].

Available: https://www.icr.org/article/cosmic-magnetic-fields-creation/%20/. [Accessed 6 November 2023].

[186] Brian Thomas, Ph.D. "Mercury's Fading Magnetic Field Fits Creation Model," Institute for Creation Research - ACTS & FACTS IMPACT, 26 October 2011. [Online]. Available: https://www.icr.org/article/mercurys-fading-magnetic-field-fits. [Accessed 6 November 2023].

[187] Russell Humphreys, Ph.D., "Impact #352 Nuclear Decay: Evidence for a Young World," Research Gate, October 2010. [Online]. Available: https://www.researchgate.net/publication/239532917_Impact_352_NUCLEAR_DECAY_EVIDENCE_FOR_A_YOUNG_WORLD. [Accessed 10 Decembrer 2023].

[188] Andrew A. Snelling, Ph.D., "Helium in Radioactive Rocks," Answers in Genesis, 8 August 2021. [Online]. Available: https://answersingenesis.org/age-of-the-earth/6-helium-in-radioactive-rocks/. [Accessed 10 December 2023].

[189] Andrew A. Snelling, Ph.D., "Helium in Radioactive Rocks," Answers in Genesis, 1 October 2012. [Online]. Available: https://answersingenesis.org/age-of-the-earth/6-helium-in-radioactive-rocks/. [Accessed 10 December 2023].

[190] "Blood," Science Museum, 10 July 2019. [Online]. Available: https://www.sciencemuseum.org.uk/objects-and-stories/medicine/blood. [Accessed 1 March 2024].

[191] E. Hovind, "Do Rabbits Chew the Cud?," Creation Today, [Online]. Available: https://creationtoday.org/do-rabbits-chew-the-cud/. [Accessed 22 August 2024].

[192] "A8 Creation of a Bacterial Cell Controlled by a Chemically Synthesized Genome45," NIH National Library of Medicine - National Center for Biotechnology Information, [Online]. Available:

https://www.ncbi.nlm.nih.gov/books/NBK84435/. [Accessed 23 January 2024].

[193] A. C. Madrigal, "To Model the Simplest Microbe in the World, You Need 128 Computers," The Atlantic, 23 July 2012. [Online]. Available: https://www.theatlantic.com/technology/archive/2012/07/to-model-the-simplest-microbe-in-the-world-you-need-128-computers/260198/. [Accessed 17 August 2024].

[194] Brian Thomas, Ph.D., "150 Years Later, Fossils Still Don't Help Darwin," Institute for Creation Research, 2 March 2009. [Online]. Available: https://www.icr.org/article/a-150-years-later-fossils-still-dont-help-darwin. [Accessed 24 October 2023].

[195] J. M. Jordan, "How many times is fear God mentioned in the Bible? (What does the Bible say about fearing God?)," Christian Faith Guide, [Online]. Available: https://christianfaithguide.com/how-many-times-is-fear-god-mentioned-in-the-bible/. [Accessed 25 October 2023].

[196] B. Porter, "The Two World Views," YouTube, [Online]. Available: https://www.youtube.com/results?search_query=Bruce+Porter+-+The+Two+World+Views. [Accessed 7 November 2023].

[197] B. Porter, "The Two World Views," YouTube, [Online]. Available: https://www.youtube.com/results?search_query=Bruce+Porter+-+The+Two+World+Views. [Accessed 7 November 2023].

[198] "Revelation 20:4-6," Bible Gateway, [Online]. Available: https://www.biblegateway.com/quicksearch/?quicksearch=reign+for+a+thousand+years&version=NIV. [Accessed 27 November 2023].

[199] "Global Conflict Tracker," cfr.org, [Online]. Available: https://www.cfr.org/global-conflict-tracker/. [Accessed 29 January 2024].

[200] A. E. Berkowitz, "2021 on Track to Become Record - Breaking year of Major Earthquakes, Volcanoes," ISRAEL365NEWS, 7 March 2021. [Online]. Available: https://www.israel365news.com/344311/2021-on-track-to-become-record-breaking-year-of-major-earthquakes-volcanoes/. [Accessed 29 January 2024].

[201] J. Duvall, "New Statistics on Growing Persecution Against Christians," Liberty Champion - The official student newspaper of Liberty University, 5 December 2022. [Online]. Available: https://www.liberty.edu/champion/2022/12/new-statistics-on-growing-persecution-against-christians/. [Accessed 29 January 2024].

[202] D. Silliman, "Decline of Christianity Shows No Signs of Stopping," Christianity Today, 13 September 2022. [Online]. Available: https://www.christianitytoday.com/news/2022/september/christian-decline-inexorable-nones-rise-pew-study.html. [Accessed 29 January 2024].

[203] "Has The Gospel Now Been Preached To All Nations?," Faith in the News, [Online]. Available: https://faithinthenews.com/has-the-gospel-now-been-preached-to-all-nations/. [Accessed 27 January 2024].

[204] "History of Jerusalem: Timeline for the History of Jerusalem (4500 BCE - Present)," Jewish Virtual Library - A Project of AICE, [Online]. Available: https://www.jewishvirtuallibrary.org/timeline-for-the-history-of-jerusalem-4500-bce-present. [Accessed 29 January 2024].

[205] A. Lamont, "Sir Isaac Newton (1642/3-1727) A Scientific Genius and "Father of Physics","

Answers in Genesis - Originally published in Creation 12, no 3 page 48-51, June 1990. [Online]. Available: https://answersingenesis.org/creation-scientists/profiles/sir-isaac-newton/. [Accessed 9 November 2023].

[206] "Top 25 Quotes by Isaac Newton (of 194) A-Z Quotes," AZ Quotes, [Online]. Available: https://www.azquotes.com/search_results.html?query=Sir+Isaac+Newton+. [Accessed 9 November 2023].

[207] D. H. M. Morris, "Sir Fancis Bacon," Answers in Genesis, [Online]. Available: https://answersingenesis.org/creation-scientists/profiles/sir-francis-bacon/. [Accessed 9 November 2023].

[208] "Top 25 Quotes by Francis Bacon (of 654) A-Z Quotes," AZ Quotes, [Online]. Available: https://www.azquotes.com/search_results.html?query=Sir+Francis+Bacon+. [Accessed 9 November 2023].

[209] Diane Severance, Ph.D., "Robert Boyle: Father of Modern Chemistry," Christianity.com, 25 January 2019. [Online]. Available: https://www.christianity.com/church/church-history/timeline/1601-1700/robert-boyle-father-of-modern-chemistry-11630103.html. [Accessed 9 November 2023].

[210] "Robert Boyle Quotes," AZ Quotes, [Online]. Available: https://www.azquotes.com/author/28103-Robert_Boyle. [Accessed 9 November 2023].

[211] Anne Marie Helmenstine, Ph.D., "Biography of John Dalton, the 'Father of Chemistry'," ThoughtCo., Updated 3 July 2019. [Online]. Available: https://www.thoughtco.com/john-dalton-biography-4042882. [Accessed 9 November 2023].

[212] "John Dalton Quotes," AZ Quotes, [Online]. Available: https://www.azquotes.com/author/26632-John_Dalton. [Accessed 9 November 2023].

[213] "Johannes Kepler Quotes," AZ Quotes, [Online]. Available: https://www.azquotes.com/author/7921-Johannes_Kepler. [Accessed 9 November 2023].

[214] "Galileo Galilei," History.com, Updated 6 June 2023. [Online]. Available: https://www.history.com/topics/inventions/galileo-galilei. [Accessed 9 November 2023].

[215] "Galileo Galilei Quotes," AZ Quotes, [Online]. Available: https://www.azquotes.com/author/5284-Galileo_Galilei. [Accessed 9 November 2023].

[216] L. J. Jean Orcibal, "Blaise Pascal - French philosopher and scientist," Britannica.com, Updated 23 October 2023. [Online]. Available: https://www.britannica.com/biography/Blaise-Pascal. [Accessed 9 November 2023].

[217] "Blaise Pascal Quotes," AZ Quotes, [Online]. Available: https://www.azquotes.com/author/11361-Blaise_Pascal. [Accessed 9 November 2023].

[218] "Gregor Mendel," Biography, Updated 21 May 2021. [Online]. Available: https://www.biography.com/scientists/gregor-mendel. [Accessed 9 November 2023].

[219] "Gregor Mendel Quotes," AZ Quotes, [Online]. Available: https://www.azquotes.com/author/26979-Gregor_Mendel. [Accessed 9 November 2023].

[220] "William Thomson," Famous Scientists - The Art of Genius, [Online]. Available: https://www.famousscientists.org/william-thomson/. [Accessed 9 November 2023].

[221] "Top 25 Quotes by Lord Kelvin (of 59) A-Z Quotes," AZ Quotes, [Online]. Available: https://www.azquotes.com/search_results.html?query=William+Thomson+. [Accessed 9 November 2023].

[222] "Werner Heisenberg," Famous Scientists - The Art of Genius, [Online]. Available: https://www.famousscientists.org/werner-heisenberg/. [Accessed 9 November 2023].

[223] "Werner Heisenberg Quotes," AZ Quotes, [Online]. Available: https://www.azquotes.com/author/6514-Werner_Heisenberg. [Accessed 9 November 2023].

[224] "Humphry Davy," Famous Scientists - The Art of Genius, [Online]. Available: https://www.famousscientists.org/humphry-davy/. [Accessed 9 November 2023].

[225] "Humphry Davy Quotes," AZ Quotes, [Online]. Available: https://www.azquotes.com/author/3746-Humphry_Davy. [Accessed 9 November 2023].

[226] "Arthur Eddington," Famous Scientists - The Art of Genius, [Online]. Available: https://www.famousscientists.org/arthur-eddington/. [Accessed 9 November 2023].

[227] "Arthur Eddington Quotes," AZ Quotes, [Online]. Available: https://www.azquotes.com/author/4345-Arthur_Eddington. [Accessed 9 November 2023].

[228] "John Ambrose Fleming," Famous Scientists - The Art of Genius, [Online]. Available: https://www.famousscientists.org/john-ambrose-fleming/. [Accessed 9 November 2023].

[229] "John Ambrose Fleming Quotes," AZ Quotes, [Online]. Available: https://www.azquotes.com/author/28168-John_Ambrose_Fleming. [Accessed 9 November 2023].

[230] "Samuel Morse," Famous Scientists - The Art of Genius, [Online]. Available:

https://www.famousscientists.org/samuel-morse/. [Accessed 9 November 2023].

[231] "Samuel Morse Quotes," AZ Quotes, [Online]. Available: https://www.azquotes.com/author/37705-Samuel_Morse. [Accessed 9 November 2023].

[232] "John Eccles," Famous Scientists - The Art of Genius, [Online]. Available: https://www.famousscientists.org/john-eccles/. [Accessed 9 November 2023].

[233] "John Eccles Quotes," AZ Quotes, [Online]. Available: https://www.azquotes.com/author/26545-John_Eccles. [Accessed 9 November 2023].

[234] "Gottfried Leibniz," Famous Scientists - The Art of Genius, [Online]. Available: https://www.famousscientists.org/gottfried-leibniz/. [Accessed 9 November 2023].

[235] "Gottfried Leibniz Quotes," AZ Quotes, [Online]. Available: https://www.azquotes.com/author/8695-Gottfried_Leibniz. [Accessed 9 November 2023].

[236] "James Clerk Maxwell," Famous Scientists - The Art of Genius, [Online]. Available: https://www.famousscientists.org/james-clerk-maxwell/. [Accessed 9 November 2023].

[237] "James Clerk Maxwell Quotes," AZ Quotes, [Online]. Available: https://www.azquotes.com/author/21477-James_Clerk_Maxwell. [Accessed 9 November 2023].

[238] "James Prescott Joule," Famous Scientists - The Art of Genius, [Online]. Available: https://www.famousscientists.org/james-prescott-joule/. [Accessed 9 November 2023].

[239] "James Prescott Joule Quotes," AZ Quotes, [Online]. Available: https://www.azquotes.com/author/23615-James_Prescott_Joule. [Accessed 9 November 2023].

[240] "Joseph Henry," Famous Scientists - The Art of Genius, [Online]. Available: https://www.famousscientists.org/joseph-henry-2/. [Accessed 9 November 2023].

[241] "Joseph Henry Quotes," AZ Quotes, [Online]. Available: https://www.azquotes.com/author/21500-Joseph_Henry. [Accessed 9 November 2023].

[242] L. H. Heydenreich, "Leonardo da Vinci - Italian artist, engineer, and scientists," Britannica.com, Updated 1 November 2023. [Online]. Available: https://www.britannica.com/biography/Leonardo-da-Vinci. [Accessed 9 November 2023].

[243] "Leonardo da Vinci Quotes," AZ Quotes, [Online]. Available: https://www.azquotes.com/author/15101-Leonardo_da_Vinci. [Accessed 9 November 2023].

[244] "Louis Pasteur - The Art of Genius," Famous Scientists, [Online]. Available: https://www.famousscientists.org/louis-pasteur/. [Accessed 13 December 2023].

[245] "Louis Pasteur Quotes," AZ Quotes, [Online]. Available: https://www.azquotes.com/author/11366-Louis_Pasteur. [Accessed 9 November 2023].

[246] "Michael Faraday," Famous Scientists - The Art of Genius, [Online]. Available: https://www.famousscientists.org/michael-faraday/. [Accessed 9 November 2023].

[247] "Michael Faraday Quotes," AZ Quotes, [Online]. Available: https://www.azquotes.com/author/4659-Michael_Faraday. [Accessed 9 November 2023].

[248] "Nicolaus Copernicus," History.com, Updated 31 January 2023. [Online]. Available: https://www.history.com/topics/inventions/nicolaus-copernicus. [Accessed 9 November 2023].

[249] "Top 25 Quotes by Nicolaus Copernicus (of 59) A-Z Quotes," AZ Quotes, [Online]. Available: https://www.azquotes.com/search_results.html?query=Nicolas+Copernicus+. [Accessed 9 November 2023].

[250] "Leonhard Euler Quotes," AZ Quotes, [Online]. Available: https://www.azquotes.com/author/4573-Leonhard_Euler. [Accessed 9 November 2023].

[251] "Arthur H. Compton - Facts," The Nobel Prize, [Online]. Available: https://www.nobelprize.org/prizes/physics/1927/compton/biographical/. [Accessed 13 December 2023].

[252] "Arthur Compton," Famous Scientists - The Art of Genius, [Online]. Available: https://www.famousscientists.org/arthur-compton/. [Accessed 13 December 2023].

[253] "Arthur Compton Quotes," AZ Quotes, [Online]. Available: https://www.azquotes.com/author/29177-Arthur_Compton. [Accessed 9 November 2023].

[254] "Charles Townes," Famous Scientists - The Art of Genius, [Online]. Available: https://www.famousscientists.org/charles-townes/. [Accessed 9 November 2023].

[255] "Charles Hard Townes Quotes," AZ Quotes, [Online]. Available: https://www.azquotes.com/author/29179-Charles_Hard_Townes. [Accessed 9 November 2023].

Author's Biography

Colonel Robert E. Johnson III, USAF, Retired

Bob is the son of Lt. Col. (retired) Robert E. Johnson II, an Air Force JAG officer and veteran of both World War II and the Korean War. Bob graduated from West Virginia University with a BS in Finance and holds an MS in Business Administration from the University of Northern Colorado. Over his 28-year career, Bob served in the following positions: Section Commander and Operations Officer at Squadron Officer School (SOS) Maxwell/Gunter AFB, AL; Chief of Supply at Hickam AFB, HI; Chief of Supply at McChord AFB, WA; Director of Supply at the Air Force Logistics Management Agency, Maxwell/Gunter AFB, AL; Logistics Group Commander at Osan AB, Korea; and Chief of Supply for Headquarters Air Combat Command, Langley AFB, VA. His Air Force awards include two Legion of Merit Medals, eight Meritorious Service Medals, two Commendation Medals, and the Department of the Air Force 1994 Dudley C. Sharp Award for Excellence in Logistics. After retiring from the Air Force, Bob worked for three defense contractors. He served as the Director of Business Development at Resource Consultants Incorporated (RCI) and for SERCO. He also served as Deputy Director and Senior Manager in the F-22 Program at Lockheed Martin, where he won the Secretary of Defense Performance Based Logistics NOVA Award for his contributions to the F-22 Product Support Integration Team. Bob is an Eagle Scout and a born-again Christian.

Made in the USA
Middletown, DE
23 August 2025

12692913R00175